ENERGY AND THE ECONOMY OF NATIONS

By the same author

THE COMMON MARKET
ENERGY IN EUROPE 1945–1980
NUCLEAR POWER

ENERGY AND THE ECONOMY OF NATIONS

W. G. JENSEN PH.D.

G. T. FOULIS & CO. LTD
50a Bell Street
Henley-on-Thames
Oxfordshire

First published 1970

ISBN 0 85429 105 9

Printed in Great Britain
at the University Printing House, Cambridge
(Brooke Crutchley, University Printer)

CONTENTS

To Anne

SECTION I

THE PLACE OF ENERGY IN THE NATIONAL ECONOMY

A fascinating, distinctly futuristic symposium was held in March 1968 in the small French town of Gif-sur-Yvette, a few miles to the south-west of Paris, to examine the effects of the development of new energy industries and the application of modern techniques in the field of energy upon existing national or regional structures. The conference was mainly concerned with developments in metropolitan France where it was expected that by the year 2020 the population would have risen to approximately 80 million (compared with some 50 million in 1967) and that the standard of living would by that time have reached a level broadly comparable to that obtaining in the United States by the turn of the century. These conclusions may also be regarded as being generally valid for the majority of advanced industrial societies in Europe and the Far East (i.e. Australia and Japan). It followed that while those countries falling into this broad category would, like France, still be some twenty or twenty-five years behind the United States, their living standards would nevertheless be some five or six times higher than they are today. (A small select group of countries like Sweden, Switzerland, Australia and Japan, will, however, by that time, have formed an intermediate group, standing somewhere between the United States and Canada, on the one hand, and Western Europe in general, on the other hand.) Production of manufactured goods, in terms of volume, was confidently expected to multiply by between eight and ten times. Raw material requirements were expected to show a six to eightfold increase; but the real key to the stupendous increase envisaged in living standards was clearly to be found in an anticipated ten to twelvefold increase in the demand for energy and power.

The forecast of soaring, wellnigh insatiable energy requirements was based on four main factors: first, vastly greater heat requirements by a broad range of industries; second, industry's rapidly growing requirements of kinetic energy—following from the steadily accelerating application of mechanization and automated techniques—which was regarded as being likely to result in a rate of increase in demand for energy that would be proportionately greater than the rise in industrial output; third, a significantly bigger *per capita* level of demand in leisure or the non-productive sectors; and fourth, a very much bigger demand for energy products as feedstock for the chemical industry.

The bulk of the increase in energy demand was expected to be for greater supplies of electric power. This, in turn, was expected to intensify and accelerate the trend that has already been so noticeable in the last two decades towards ever larger generating stations. This was particularly the case for nuclear power stations where big savings in capital costs were envisaged as a result of very much bigger unit sizes. It was at this stage that the participants in the symposium, for once discarding some of their customary caution, peered into a future teeming with exciting, if not also a little terrifying, technological developments. It was in this frame of mind that the conference listened to papers outlining the creation of vast nuclear complexes of some 2,000 to 3,000 MW apiece, each one capable of producing up to sixty billion kWh a year—equal to almost 30 per cent of the total output of electricity in the United Kingdom in 1966. Electric current would be transported from these generating centres to the main areas of consumption by means of giant new super high-tension lines of 730 kV. Since power packs of this size and complexity would create enormous demands for water for cooling purposes they would perforce have to be constructed on or near the coast, where a proportion of the power they produced could easily be earmarked for desalination purposes. Even more exciting and seemingly akin to the realm of science fiction was the suggestion that the problem arising from the accumulation of atomic waste that must inevitably result from the widespread construction of nuclear power stations could be satisfactorily resolved by its expulsion, in special containers, by rockets despatched on trajectories of non-return into the remoteness of outer space.

We shall have occasion to come back again, in a later Section, to the conference of Gif-sur-Yvette; but before doing so we must proceed to somewhat more mundane issues and consider first, briefly, the exactness and reliability of energy demand forcasting and, secondly, the crucial question of the place and role of energy in the national economy.

The science of energy demand forecasting

Since 1945, energy forecasting has become a science with its own highly sophisticated techniques. It has nonetheless remained a hazardous occupation, subject to a large number of both internal and external imponderables. Internal factors of uncertainty include the inescapable need to proceed on what must of necessity be a number of basic but unproven assumptions, the often inadequate availability of past data, and the existence of organizational or procedural difficulties regarding accessibility to current experience or climates of opinion that are capable of exerting strong influences on the development of economic growth. Many of these internal factors can, in the final analysis, be identified as problems of communication and their solution is in many respects an acknowledged test of the successful co-ordination of governmental, industrial and technical expertise. External factors, such as climatic conditions, fluctuations in the price of energy imported from outside national frontiers or regional groupings, as well as changes in overall energy availability as a result of localized political disturbances, on the other hand, constitute areas where the ability of governments to insure against the unknown or unsuspected is very much reduced.

In the preparation of energy forecasts countries may, broadly speaking, be divided into three main categories. First, there are the states—such as the so-called socialist economies of Eastern Europe—with centrally planned economies, where in the majority of cases a central planning commission co-ordinates the plans prepared by the responsible ministries. These plans are generally for five-year periods, with yearly comparisons and a consequent revision for the remainder of the period in question. They may be supplemented by 15–20 year-period perspective plans or projections which serve to more clearly identify long-term objectives. The second category is made up of countries, such as France, where centrally initiated national plans—usually covering periods of five to six years' duration with associated long-term projects—are drawn up which, while not obligatory in their application, are nevertheless intended to encourage the private sector so to shape its forward planning, investment and development as to keep in step and in harmony with progress in the key sectors of the national economy under public control. There is, finally, a third and very much more heterogeneous group of countries without any kind of regular or systematic planning procedures. Countries in this group tend to rely for their energy forecasts on periodic forward studies by various Government departments, committees of experts drawn from industry or universities, forecasts by major utility or other energy supply organizations, and *ad hoc* investigations by accredited public or private bodies.

A recent United Nations survey of energy forecasting techniques concluded that

the accuracy of a projection is not in itself sufficient evidence of a good forecasting practice. In unplanned or market economies, causes of fluctuations in a particular year, or alterations in trends of activity, may sometimes lead to fortuitous accuracy in the terminal year. Moreover, aggregations of total energy consumption may well arrive at the forecast result despite important errors in the forecasting of changes of structure within different supply sectors.

Neither does it follow that complex methods of projection are necessarily more successful than simple methods. What is more important is that

—the methods chosen are well adapted to the tendencies and characteristics of the form of energy or consuming sector to which the projection refers;

—the length of the forecast and arrangement for regular revision are adapted to the purpose in hand;

—access is sufficiently assured to information as to possible changes in the conditions affecting requirements.

Some recent experience has been salutary in this connexion. Thus, the electricity projection made in France for the Fourth Plan (covering the five-year period from 1961 to 1965) was based on formulae which allow fully for the planned economic growth and take account of probability considerations. Nevertheless, the discrepancy at the end of the planning period amounted to 6·6 per cent whereas a straightforward extrapolation would have proved correct to within—4·5 per cent. In practice this departure was able to be largely corrected before the end of the plan period by successive adjustments which took the changed situation into account. In some countries, such as Portugal, similar experience has led to simple extrapolations being preferred to more elaborate methods in dealing with electric energy.

Turning to measures now being taken or developed with a view to promoting what it describes as predictive efficiency, the report continued:

Comparison between different plans and forecasts offers some guidance as to the positive steps which can be taken to create conditions likely to be most clearly adapted to the purpose in hand. For comprehensive projections in market economies the best results appear to be reached in dealing with the problem of competitive fuels where the projections are made regularly at frequent intervals, with as short a period of coverage as is consistent with the objectives, and with more provisional but regularly-revised extrapolations added to extend beyond the operative period of the forecast. Since considerable errors have tended to occur in predicting the trend of economic activity where no economic plan exists it may also be concluded that this kind of approach is more likely to produce useful results where it is embodied in a comprehensive national plan. An alternative approach—that of the medium-term plan and associated long-term projection with the former compared annually with achievement and revised accordingly—is that used in countries with a centrally-planned economy and also in market economies such as that of France. This method allows more elaborate investigations to be made at the outset and can give good results under market conditions when the problem of competitive fuels has been more clearly resolved. In any event, much depends on the availability of a long and reliable statistical basis, preferably assembled as a time series and also in the form of annual balances for at least a representative series of years.

It is difficult to conclude, from evidence available so far, that econometric models give superior results for the purpose of projecting energy requirements.

On the other hand, the use of some formulae of an econometric type has given good results in projecting trends for particular forms of energy such as electricity.

It is worth pausing for a moment at this stage to look at two or three examples of the impact made by policy considerations in determining energy forcasts and the methods used in market analysis. For our first example we shall look at a study made a few years ago in France in order to determine the probable breakdown by types of fuel of the energy supplies needed to achieve the anticipated requirements of thermal electricity. First there were fed into a model elaborated for this purpose two alternative sets of figures expressing the ratio between prices of oil and coal units of equal calorific value. The country was split up into a number of regions and an assessment made of the total fuel requirements of their thermal plants. Next, calculations were made, on the basis of these assumptions, of the prices at which both coal and oil supplies could be delivered at different periods in the future. The picture that then emerged was used as a basis for estimating the future pattern of fuel usage at power stations in each individual region including, for example, indications of the points in time when only oil or nuclear stations would be built or when it might be economically viable to convert coal-fired plants to oil or natural gas. Only then were the quantities of coal, oil or other fuels determined that would be required for electricity generation both by regions and by the country as a whole. The end product was, economically, fascinating. According to whether the price of oil fell in relation to that of coal by one tenth or one fifth between 1960 and 1975, coal's share of the French power generation market was expected to decline by 27 or 65 per cent.

	Anticipated percentage of coal-firing at power plants over the period			
Oil/coal price ratio assumptions	1959	1965	1970	1975
Fall of 10 % in the price of oil in relation to coal	80	80	65	53
Fall of 20 % in the price of oil in relation to coal	80	75	41	15

But while the application of the pure economic doctrine underlying the calculation was faultlessly executed, the reality of the situation has been very different. Social, financial and regional considerations have obliged the French government to intervene in order to increase the level of coal consumption well above the levels dictated by purely economic, or rather arithmetical, calculations.

The impact of long-term policy considerations upon energy forecasting

was clearly demonstrated in the British White Paper on Fuel Policy of November 1967, where the production or consumption figures given for the various sources of energy in 1975, in some cases very different from the estimates submitted by the individual fuel industries, were the outward sign of the overall pattern of energy development that was anticipated and, presumably, desired by the British Government. A glance at the various factors which the Government departments took into account in the formulation of these figures shows at once how many of them are subject to governmental control and manipulation:

(i) the quantity of natural gas likely to become available from the North Sea; the most economic method of its absorption; and its price to the gas industry;

(ii) the relative selling prices, including transport and other charges where appropriate, of the various fuels;

(iii) the rate of development of nuclear power;

(iv) the rate and pattern of growth of the economy;

(v) the level and pattern of protection for indigenous fuels.

On points (i) and (iii) to (v) the Government's power of intervention is or can be decisive. Clearly, therefore, the final shape of any forward energy projection must and will be based upon Government views in such matters as the optimum rate of development of nuclear power and nuclear know-how (albeit for reasons not immediately related to cheap energy), taxation and fiscal policy, or, conversely, a propensity for import saving. In brief, the decisions determining the future levels of consumption of different forms of energy become politically, and no longer purely economically or commercially, motivated.

Whereas the previous example related to a largely coal-based national energy supply with little or no hydropower but where nuclear power is being introduced on a very substantial scale, our third and last example in this field concerns the incorporation of nuclear power into a system where hydro-power has traditionally supplied the lion's share of total energy output. For it, we must go back to 1962 when a study was made in Sweden of the probable structure of electricity supply during the 1970s to determine the most economic allocation of production between various sources of energy. The study was thorough and wide-ranging, allowing for three alternative hydro developments as well as several combinations of thermal and nuclear power plants, two alternatives for oil prices and three possible levels of nuclear costs. From the analysis it emerged that the degree of the variations allowed for in the price of oil or in fixed costs of nuclear power were unlikely to affect the optimum balance between hydro and thermal power to any significant extent since the difference in incremental costs remained slight. Such a situation only applied, however, to hydro systems, such as the one in Sweden, where the unit development costs of unused resources were fast approaching the exploitable limit. While it was accepted that in a comparison of this kind the basic assumptions, which were clearly of fundamental importance, were unproved and of a hypothetical character, the study was nevertheless regarded by the Swedish authorities as having demonstrated that, within the range of the parameters used, the realization of favourable conditions for the development of nuclear power would make it a highly attractive proposition, whereas unfavourable conditions would not make it unduly unattractive or economically disadvantageous. The study, in short, was looked upon as constituting a reasonable basis for judgement despite the continuing and unresolved uncertainties about fuel prices and nuclear costs.

The growing importance attached to the exactness and reliability of energy forcasting is, of course, a reflection of the important part energy plays in the economy of nations—and in particular in the entire way of life of an industrial society. It is to the relationship between energy consumption and economic growth that we must now turn.

Energy consumption and economic growth

Between them coal, oil, natural gas, hydro-power and nuclear energy provide the power that is needed in ever-increasing quantities by both industrial and domestic consumers. It is a sobering thought that a Saturn V rocket launching consumes as much energy in two minutes as was used to build the Great Pyramid of Eygpt in twenty years. More optimistic, perhaps, is the thought that once the energy of the oceans can be harnessed by fusing the heavy hydrogen isotope deuterium into helium (and nuclear scientists are convinced that this is simply a question of time), then two cubic kilometres of water will yield as much energy as has been used by man since the beginning of time up to the present day.

The simile likening modern energy to the slave labour of past ages is, therefore, not altogether inappropriate, with this important difference that energy today is no longer the preserve of a favoured few but is at the service of society as a whole. Without adequate supplies of energy, our industrial world would soon be paralysed. We have only to reflect for a moment upon the part played by electricity, oil or coal in our daily lives at home, on the road, in the factories, offices or other places of work to realize the pervasive importance and significance of energy. Alternatively, we have only to cast back our minds to the situation that prevailed over

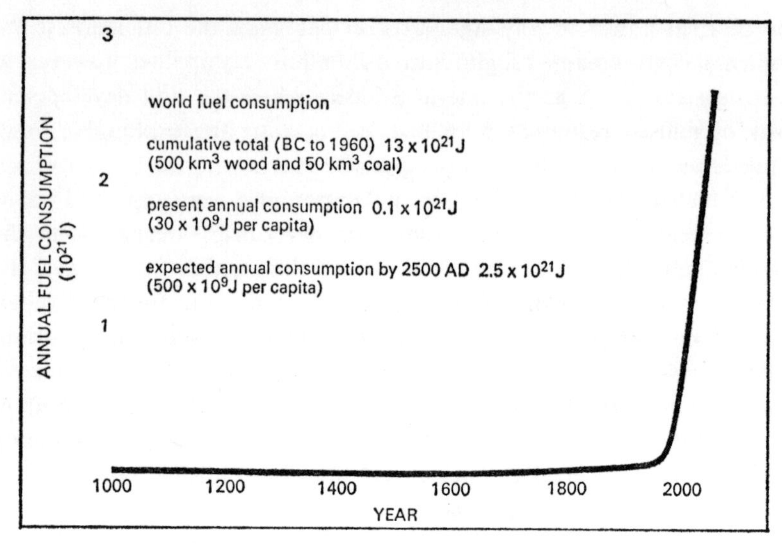

Figure 1. Growth of world energy consumption
Source: Science Journal, January 1968, p. 36

much of Europe, including Britain, and Asia in 1945 to get a very clear picture of the grave problems that arose at that time from a situation of near universal energy shortage. An American report, published in October 1947, stated notably that

countries in Europe have been obliged to continue restricting the household and non-essential industrial consumption and to concentrate the major part of their available supplies on essential industries. In most countries, the allocation of coal was and still is based on a system of priorities. In general, first priority is granted to the requirements of public utilities (gas, electricity), and transport. Next in order of importance come the essential industries such as iron and steel, food and (in many cases) building materials. According to local circumstances and weather conditions, priority is also granted in some countries to domestic consumption over non-essential industries. It follows from this that, in periods of acute shortage such as occur during the winter owing to transport difficulties, industrial activity is seriously curtailed and the progress of national recovery impeded. For example, in France and Denmark it proved necessary to order a complete stoppage of industrial activity for two or three weeks in the middle of the winters of 1945/46 and 1946/47 save in the case of continuous processes. In other countries the stoppage was confined to certain industries, e.g. textiles in Belgium, or areas, e.g. south-eastern England when stocks at power stations ran out. Conditions such as these have borne with particular weight on those countries that have little or no indigenous production and have been alleviated only in part by the system of international allocation designed to assure at least minimum requirements of all participating countries being met.[1]

[1] US Department of State, Washington, Publication 2952: Committee of European Co-operation (European Series 29), published October 1947, Vol. II, p. 141.

In some cases there was real hardship as a result of the fuel shortage and the winters of 1944/45, 1945/46 and 1946/47 saw many people in countries like Belgium, Germany and Holland breaking down the doors of their rooms, furniture and even window frames in little used rooms, to provide some kind of fuel for their fires.

When, by the early 1950s, Europe, as a result in large measure of American Marshall aid, was well on the way to recovery, the anticipated continuing shortage of energy supplies constituted one of the major economic issues of the day. It was precisely this state of shortage that gave the initial impulse to the development of the first European nuclear power stations, particularly in the United Kingdom, despite the fact that electricity generating costs at these stations seemed likely to remain, for a considerable period to come, well in excess of those at conventionally fuelled plants.

The same conviction that energy was likely to remain in short supply led to European Governments drawing up plans for big increases in their indigenous coal production, the signature of massive coal import contracts from the United States and a boost to the overseas exploration programmes of both American and European oil companies. Above all, Governments, in the face of the mounting evidence of the intimate relationship between rising demand for energy and industrial growth, continued to regard the availability of energy, particularly from indigenous sources, as a major policy matter.

While the extent of the closeness of the relationship between industrial and economic growth on the one hand and energy demand on the other hand has become an accepted tenet of economic faith, nowhere has this fact been more amply illustrated than in the voluminous supporting documents prepared by the responsible agencies of the European Communities in the course of their attempts to formulate a common energy policy for the six Common Market countries.

The Community experts worked on the figures (1950–60) and assumptions (1961–75) with regard to annual percentage increases in Gross National Product, industrial production and energy consumption shown in Tables 1–3.

A feature of the whole of the period between 1950 and 1975—and indeed as far ahead as existing projections permit one to see—in the Common Market countries as in other industrialized parts of the world, is the particularly rapid growth that has taken place and is expected to continue to do so in the demand for electric power. The extent of the difference in the relationship between increased Gross National Product and growth in demand for electricity and overall energy requirements is

Table 1. *Increase in G.N.P. in the Common Market Countries 1950–75*[1]

	Actual	Estimated		
	1950–60	1960–65	1965–70	1970–75
Belgium	2·7	3·8	3·9	3·9
France	4·3	5·2	4·7	4·6
Germany	7·4	4·4	4·0	4·2
Italy	5·9	5·95	5·75	5·3
Netherlands	4·9	4·3	4·9	4·3
Community	5·5	4·9	4·6	4·6

Table 2. *Increase in industrial production in the Common Market countries 1950–75*[2]

	Actual	Estimated		
	1950–60	1960–65	1965–70	1970–75
Belgium	3·0	4·8	4·8	4·8
France	6·4	6·5	5·9	5·5
Germany	9·1	5·5	5·0	5·0
Italy	8·1	8·8	7·8	6·5
Netherlands	5·8	5·4	6·0	5·6
Community	7·5	6·3	6·0	5·6

Table 3. *Increase in energy consumption in the Common Market countries 1950–75*[3]

	Actual		Anticipated		
	1950–60	1955–60	1960–65	1965–70	1970–75
Belgium	3·4	0·3	1·8	2·7	2·5
France	4·4	3·6	4·4	4·3	4·3
Germany	7·0	2·7	3·0	3·4	3·2
Italy	10·4	7·8	8·6	6·6	5·2
Netherlands	4·7	3·6	4·9	3·8	3·8
Community	6·1	3·5	4·3	4·2	4·2

well illustrated in Figure 2: whereas, taking the period from 1950 to 1975 as a whole, Gross National Product was expected to increase nearly fourfold and total energy demand to rise by a factor of three, electricity requirements were expected to rise no less than sevenfold.

[1] Étude sur les perspectives énergétiques à long terme de la Communauté européenne, Luxembourg, 1964, p. 19.

[2] *Ibid.*, p. 20.

[3] *Ibid.*, p. 20.

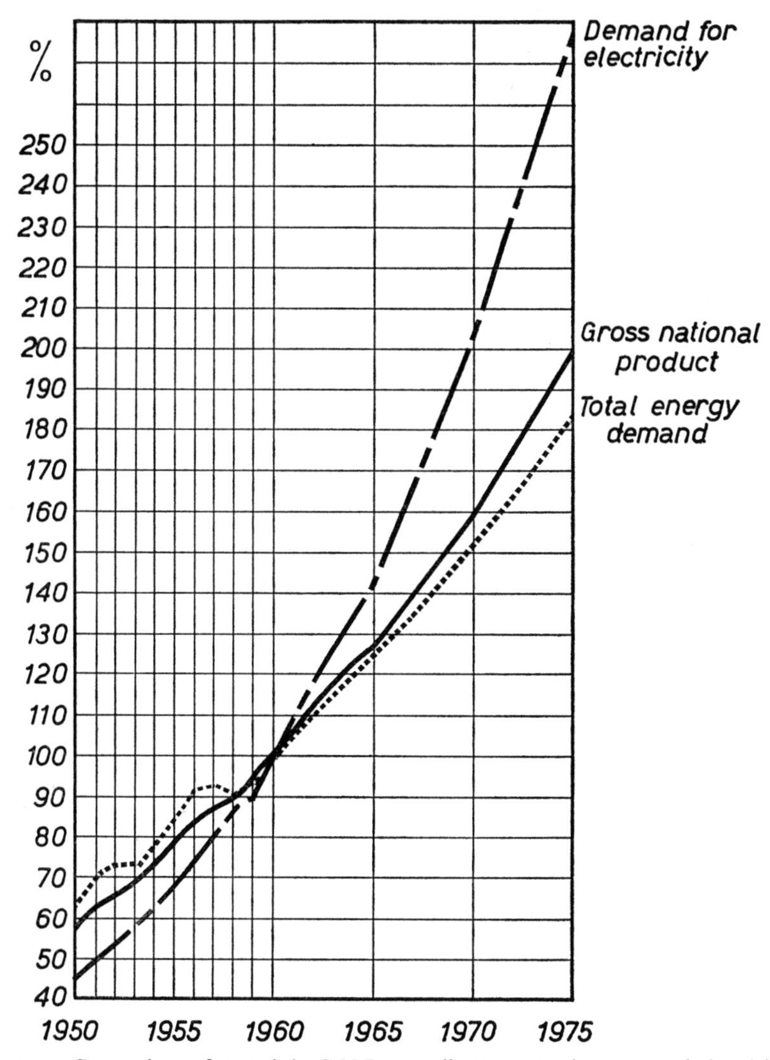

Figure 2. Comparison of growth in G.N.P. overall energy requirements and electricity demand in the six Common Market countries between 1950 and 1975

In terms of cold, harsh figures for the Community area as a whole, between 1950 and 1960, an average yearly increase in Gross National Product of 5·5 per cent was paralleled by an increase in industrial production of 6·9 per cent and a rise in energy demand of 4·8 per cent. The ratio of increased energy demand to increased Gross National Product was therefore of the order of 0·87—or considerably less than parity. During the period from 1960 to 1965 the annual rate of increase in Gross

National Product, at 4·9 per cent, was expected to be about 10 per cent lower than in the previous decade, while the corresponding figures for industrial production and energy requirements, at 6·3 per cent and 4·3 per cent, were expected to show broadly similar reductions. The ratio of increased energy demand to increased Gross National Product was accordingly expected to remain virtually unchanged at just under 0·89 per cent. In the event, the rates of increase in both Gross National Product and energy demand during this period averaged 5 per cent a year. There were signs, moreover, that for the period from 1965 to 1970, and beyond, this ratio would rise steadily in favour of energy consumption to a little above parity as the rapid expansion of the high energy-consuming industries i.e. aluminium smelting, electro-mechanical and electro-metallurgical plants) and demand for greater comfort both at home (i.e. central heating, electrical household equipment) and outside forced up the total of global energy consumption. Expressed in a different way, there was an underlying implication that the failure to obtain the necessary increase in energy requirements would seriously affect countries' ability to achieve their desired rate of increase in Gross National Product and, consequently, in national prosperity and wellbeing. Already by the 1950s a high level of *per capita* energy consumption had become an accepted and recognized yardstick to measure a given country's degree of industrialization. (See Figure 3.) By 1960 the average *per capita* consumption in the Common Market countries was 2,726 kg, compared with 4,800 kg in the United Kingdom and 8,200 kg in the United States. By 1966 these figures had risen to 3,501, 5,139 and 9,595 kg respectively, while the estimates for 1975, at 4,500, 5,690 and 10,810 kg were indicative of the vital contribution made by energy to the scientific and technological processes of contemporary society. Above all, the rising tide of energy demand is symptomatic of the process of industrialization and of the affluent society. Indeed, it is possible almost to subdivide countries into their several categories on the basis of their *per capita* consumption of energy. Below 1,000 kg are the underdeveloped countries; between 1,000 and 2,000 kg those at or beyond 'take-off' point; above 2,000 kg are the industrially advanced countries; and, lastly, above 3,000 kg come the leading countries of the technological age, headed, inevitably, by the United States, but including also some much smaller countries (in terms of population) such as Canada, Luxembourg and Sweden. The importance of energy to our everyday comfort and convenience was given a particularly dramatic demonstration by the power blackout that occurred along the north-eastern seaboard of the United States in 1965 and, to a lesser extent, by the power cuts that were for so many years after the war a regular feature of the British electricity industry.

The importance of energy and of energy costs has been recognized in all industrial countries. We have only to turn to the recent successive White Papers on this subject in the United Kingdom, to the detailed examinations prepared by the European Communities, as well as by the European

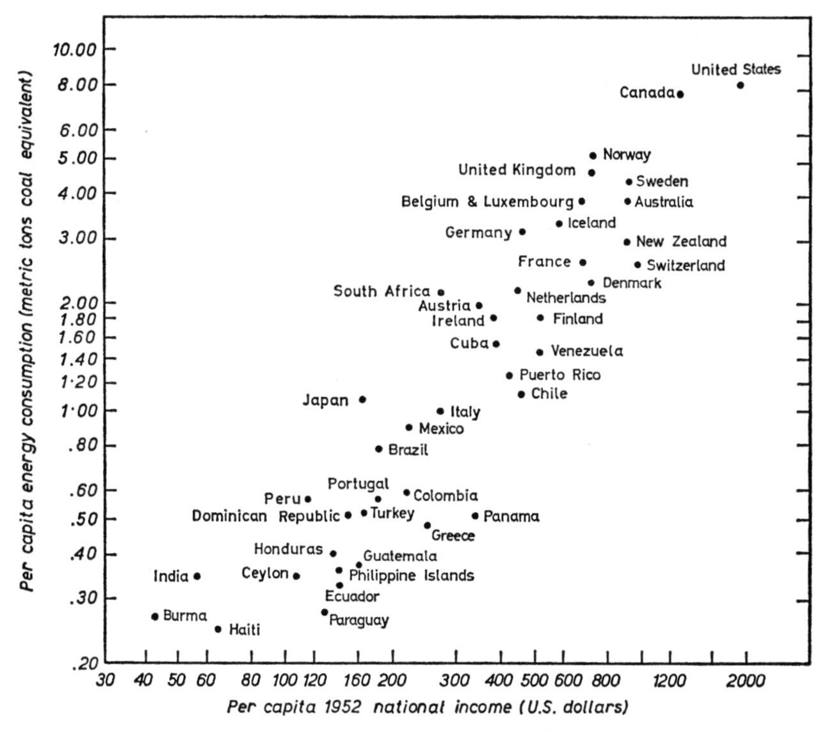

Figure 3. Energy consumption *vs.* national income, 1952
Source: U.N. Statistical Office

Parliament and the Council of Europe in Strasbourg. A report published in October 1963 by the Energy Committee of the European Parliament, for example, gave the following objectives for a Community energy policy:

(i) Ensuring adequate energy supplies at the cheapest possible price;
(ii) Security of energy supplies;
(iii) Avoidance of any major social or market distortions as new sources of energy are substituted for old sources;
(iv) Long-term supply stability;
(v) Free choice for the consumer.

The sixth objective, the maintenance of the cohesion of the Common

Market, had a purely Community application. Subsequent reports by the Energy Committee and the European Commission added two further objectives: first, the need to avoid social tensions or the emergence of an excessive disequilibrium between neighbouring regions as a result of the introduction of new forms of energy; and, second, the need to provide adequate job and salary protection for those employed in a contracting energy industry. The co-ordination of the Community's energy policy is one of the major preoccupations of the European Commission, although the conflict of interests among the six Member States has so far prevented, despite ten years of assiduous preparation, any tangible or meaningful progress. Similar considerations have loomed large in the minds of American and many other continental European Governments and have been the subject of three exhaustive reviews by the Organization for Economic Co-operation and Development (O.E.C.D.) in Paris.

There are, of course, several facets to the energy problem. Thus, in addition to the generally recognized need to secure the regular availability of energy supplies, there are the problems arising from the competitive interplay of different forms of energy as well as from indigenous and imported sources. The battle that has been waged during the last decade between coal and oil, in particular, led in turn to new problems that are at once regional, social and financial in their context and impact. There are at the present time, in many countries, vast subsidies being paid from the national exchequer to assist and maintain the coal industry. At the same time the size of the second nuclear programme in the United Kingdom is evidence of the importance attached, for instance, by the British Government to the medium term goal of securing a new source of ultimately cheap and essentially indigenous energy, since such a programme cannot be justified on grounds of current competitive cost or possible shortages of conventional energy supplies alone. An analytical appreciation of all the costs incurred would appear to be the obvious and most desirable solution to sort out the conflicting claims of coal, oil and nuclear power. But at once a whole host of difficult problems arise. How are nuclear costs to be assessed? How should the extremely expensive nuclear development costs be covered? Should costs of this kind be assigned to the military or civil budget, and, if the latter, directly to the public purse or to the charge of the electricity authorities concerned? Where countries face a shortage of capital, decisions must be taken on national priorities, since the building of a nuclear power station inevitably involves high capital expenditure and may thus be a severe strain on countries' financial resources. The pricing policy pursued by the British Gas Council with regard to natural gas supplies from the North Sea points to a similarly commendable objective of securing cheap energy supplies. The same preoccupation can be quite clearly detected in many continental European countries. Because of it the Dutch have determined on the closure of their coal industry and a similar reasoning lies behind the massive cutback in coal output in Belgium, France and Germany. But of course energy cannot be treated in isolation. In a world of growing interdependence of peoples, economies and industries, energy policy decisions cannot be extricated from problems of regional development, manpower redeployment and financial policy. In the United States, on the other hand, the question of whether to allow cheap oil imports from outside sources, such as the Middle East or North Africa, which have hitherto been strictly limited, is essentially a military and strategical one: the point at issue being whether the United States should allow itself, for the sake of passing temporary economic advantages, to be placed in a situation where it would be unable to meet its oil requirements from its own indigenous resources in an emergency. We shall come back to these complex and closely interrelated factors in Sections III, IV and V of this book.

The share of energy in the Gross National Product of the Community countries was put, in a study commissioned by the European Commission and published in 1966,[1] at about 5 per cent—ranging from 6·8 per cent in Belgium to 3·6 per cent in Italy. When calculated on the basis of energy's share of the contribution made to Gross National Product by industrial production, the result was a mean figure not far short of 11 per cent, ranging from 14·9 per cent in Belgium to 9·1 per cent in Italy. The Community study looked at the cost of energy both for the economy as a whole and for a number of different sectors; energy costs being defined as the actual costs paid by end-use consumers regardless of whether the type of energy in question was used for domestic, commercial or industrial purposes. By using this method, it was claimed, it was possible to establish energy costs for the economy as a whole by multiplying the final consumption figures by the prices paid. In case some of these percentage figures seem small, it may be useful to stress once again how vital energy availability and energy costs are to the whole process of industrial location, industrial production and industrial competitiveness. On the one hand there is a fundamental need for energy, as the irreplaceable life-blood of our modern, technical and industrial society; and, on the other hand, the determining influence that energy costs have exercised in the past, and are still exercising today, on the problem of industrial location as well as

[1] L'Influence économique du prix de l'énergie, Brussels, 1966.

on the ability of certain, often more traditional industries to sustain their competitive power and position.

Other, more detailed, figures which emerged from the Community study showed that solid fuel costs to end-users accounted for only 4 per cent of total energy costs in Italy but up to 25 per cent in Belgium and Germany; for oil and natural gas, on the other hand, the corresponding figures were 50 per cent in France, Italy and the Netherlands, 40 per cent in Belgium and approximately one-third in Germany; while those for electric current came to about 25 per cent in Germany and Italy and between 13 per cent and 17 per cent in the three other Community countries. These figures, it should be noted, related to the early years of this decade. Corresponding figures for 1967/68 could be expected to show a fall of up to one-third for solid fuels, and compensating increases for electric current, oil and natural gas. An interesting feature that was brought out with striking effect in the Community study was the very great extent of the Common Market countries' reliance upon energy imports.

Developments in individual markets

The increase in energy demand, though common to all industrialized countries, has varied in extent, nature and rapidity both between countries and individual regions. A feature of them all, however, has been the extraordinarily rapid rise in the demand for electricity. A report published in 1966 by the Energy Committee of the Organization for Economic Co-operation and Development (O.E.C.D.) in Paris[1] broke down demand by main consumer sectors in North America, Western Europe and Japan for the period 1950 to 1964, together with forecasts for the years 1970 and 1980. Since this Committee consists of senior Governmental representatives, the figures published in this report can, and indeed must, clearly be taken as reflecting the official forecasts of the member countries of the organization (i.e. the United States, Canada, the Common Market countries, the United Kingdom and most of the other European Free Trade Association's member countries, as well as Greece, Iceland and Turkey).

The O.E.C.D. report showed that in North America (i.e. U.S.A. and Canada) final consumption of energy (i.e. excluding transformation losses and non-energy products such as hydrocarbon feedstocks for the chemical industry) rose from 1,047 million tons of coal equivalent in 1950 to 1,572 million tons in 1964 and were expected to rise further to 2,685 million tons by 1980. Corresponding figures for energy requirements by

[1] Energy Policy: Problems and Objections, O.E.C.D., Paris, 1966.

the electricity generating industry were 178, 403 and 938 million tons of coal equivalent.

In the O.E.C.D. area it was similarly the electricity generating sector which has shown the most rapid rate of increase in demand: from 114 million tons of coal equivalent in 1950 to 264 million tons in 1964 and a forecast figure of 657 million tons in 1980. Total final consumption figures for the same three years were given at 504, 828 and 1,540 million tons of coal equivalent.

Most striking of all, however, was the rate of increase in energy demand for electricity generating purposes in Japan, i.e. from an admittedly low figure of 12 million tons of coal equivalent in 1950 to 57 million tons in 1964 and a forecast figure of 217 million tons for 1980: equivalent to a near twentyfold increase over the 30-year period. This compares with rates of increase of some 530 and 600 per cent in North America and Europe respectively. Total final consumption figures in Japan for 1950, 1964 and 1980 were given as 41, 148 and 442 million tons of coal equivalent respectively.

The energy market can be conveniently divided into two main parts: first, those specific markets whose requirements can only be met by one particular type of fuel; and, secondly, those markets for which all forms of energy are suitable and where consumer choice is consequently determined by price and convenience of handling. Examples of the first category are the transport market which today depends almost exclusively on oil, and the steel industry which, until such time as direct reduction becomes technically and commercially viable, remains largely dependent upon coal with satisfactory coking characteristics. General industry, the power stations and domestic consumers on the other hand, fall into the second category since coal, oil, natural gas and electricity can in most cases be used indiscriminately.

Although price remains the prime element in determining consumer preference for any particular type of fuel, it must of necessity remain subject to a reasonable assurance of continuity of supply. The 1956 Suez crisis which forced up the price of oil and severely restricted oil availability in a number of countries for a period of some months delayed the full impact of a changed market situation in Europe and elsewhere, to be characterized by cheap and plentiful oil supplies, by the best part of two years. The 1967 Middle East war, although much more limited in its effect upon oil supplies (despite the prolonged closure of the Suez Canal), also led to a small but nonetheless perceptible rise in oil prices in a number of Western countries. Even so, both industrial and domestic consumers tend, inevitably, to take a comparatively short or, at most, medium term view with regard to energy prices and prospects. It is, for example, difficult for an industrial enterprise, confronted with keen competition from its competitors, to ignore the potential—even if possibly temporary—benefits of an offer of cheap oil supplies. It is left to governments to concern themselves with the long term problems and responsibilities that arise from security of supply or social and regional difficulties resulting from the closure of uncompetitive indigenous sources of production which find their products undercut by cheaper imported energy supplies.

In the final analysis the two determining factors in any given country's attitude to the problems that arise in connection with energy policy formulation are, inevitably, availability and prices. Understandably, a country such as Italy, with virtually no indigenous sources of supply, is concerned solely with obtaining overseas supplies at the cheapest possible price. Countries like Britain, France and Germany, on the other hand, with large indigenous coal industries, producing fuel at a cost that may at times be well above the price at which alternative sources can be imported, have to weigh costs arising from regional reconstruction problems, additional import bills and the loss of an element of security, against the commercial advantages accruing from cheaper energy imported from overseas. On the other hand, we have already seen how the United States Government, despite a wealth of indigenous resources, has consistently followed a policy of restricting cheap oil imports from overseas in order to maintain a large and viable indigenous oil industry for essentially military and strategic considerations—even though the strength and influence of the American oil lobby should not and indeed cannot be overlooked. The action taken by the United States government must, however, be kept in perspective: American energy prices generally are well below those prevailing in Europe or Japan.

The effect of the transport revolution on energy availability

The tremendous developments that have taken place in the last ten to fifteen years in the field of energy transportation have been of fundamental importance in radically altering the world's global energy situation.

These transport developments cover the growth and spread of the pipeline system, the building of larger ships to carry oil and coal, the growing use of liner trains and the introduction of new methods of transporting natural gas in the liquid state. In 1945 there were only a few pipelines in operation outside the United States. Even by the early 1950s such schemes as the Bechtel plan to bring crude oil by pipline from the

Middle East to Europe were considered excessively ambitious. Today, Western Europe, in particular, has developed a dense network of pipelines for crude oil, refined oil products and natural gas, running from Rotterdam, Marseilles, Genoa and Wilhemshaven to major inland consumer areas. A new pipeline network is fast radiating out from the big natural gasfield at Groningen in north-east Holland to a number of neighbouring European countries and plans have been put forward for the construction of a giant pipeline to transport up to six thousand million cubic metres a year of natural gas from the Ukraine to Italy and France. Plans have even been mooted to carry oil and natural gas across the Mediterranean from North Africa to southern Europe. The costs of construction are often high, but these are rapidly offset by the subsequent reductions in transport costs. More recently, work has been progressing at an impressive pace to lay pipes to enable the natural gas deposits of the British North Sea area to be brought ashore and connected to the national grid.

Even more spectacular, perhaps, than the development of the European pipeline network has been the increase in size of oil tankers and, on a more modest scale, in coal-carrying ships. In the immediate post-war years the largest available tankers were of the order of 15 to 20,000 tons. The launching of new tankers of 35,000 tons was a major event in the shipbuilding world and, as recently as 1956/57, only some twelve years ago, ships of 50 to 60,000 tons were regarded as being of the supertanker class. Today, some twenty ships of 200,000 tons are either afloat or under construction. The Gulf Oil Company's Bantry Bay project envisaged the construction and chartering of six tankers of some 300,000 tons capacity—some of which are already in service. These giants bringing oil from the Middle East to Bantry Bay on the southern coast of Ireland, where their cargoes are transhipped into smaller vessels for shipment to the refineries of northern European countries. The Gulf Company intends, moreover, to build two more of these enormous terminals at Okinawa in the north-western Pacific and at Point Tupper in Nova Scotia. Some idea of the savings achieved by the introduction of these giant vessels and terminal facilities was given by Mr Binsted, the Gulf Oil Company's worldwide co-ordinator of transportation, at the 1968 Europoort Congress in Amsterdam. Mr Binsted then said:

> Although...it will be some time before we are able to chronicle the precise economics, we certainly are confident that the cost per barrel of bringing oil from Kuwait by the mammoth tankers is roughly one half as much as Gulf's cost of moving the oil through the Suez Canal in the largest vessels that could transit the Canal prior to its closure.

Within the last twelve to eighteen months Japanese shipyards have begun feasibility studies of tankers of 500,000 and one million tons capacity; and, indeed, in April 1969, it was announced that the Japanese Tokyo Tanker Company had taken a definite decision to proceed with plans to build a 450,000 ton ship. At about the same time unofficial estimates put the number of tankers in the super-giant class, of 200,000 tons and above, that would be in service on the world's trade routes by the early 1970s at over two hundred.

One major result of these developments has, of course, been to throw serious doubts on the future economic and financial viability of the Suez Canal when it is eventually reopened. In 1966, its last full year of operation, 166·7 million tons of oil passed through the Suez Canal on its way to Europe and only about 10 per cent of Europe's oil supplies from the Middle East took the long route round the Cape. Today, it has been estimated by Shell that some 45 per cent of world tanker capacity would be unable to pass, when laden, through the Canal at its present depth, and this proportion is expected to rise to nearly 60 per cent by 1971. While it would be physically and technically possible to deepen the Suez Canal so as to give it sufficient draft to take tankers of up to 250,000 tons at least when they are in ballast, such an operation would inevitably require a great deal of capital expenditure. With Egyptian resources already strained to the limit, money would clearly have to be put up by foreign sources. The world's major oil companies are, however, hardly likely to resume shipments through the Canal—assuming political stability has first returned to the area—unless transit dues can be pitched at sufficiently low and attractive levels. This, in turn, however, would throw serious doubts over the Suez Canal Authority's ability to obtain adequate returns upon a capital investment of this magnitude.

This trend towards ever larger ships has also been applied, although on a significantly smaller scale, in the coal-carrying trade, where the French, in particular, have recently constructed two vessels of 65,000 and 80,000 tons respectively, to carry coal from Hampton Roads in the United States to the new Le Havre power station and the Dunkirk steelworks. It is noteworthy also that both the United Kingdom and Poland are at present engaged in the process of constructing new coal-loading terminals, at Immingham and Swinoujscie (the old German port of Swinemunde) respectively, which will be able to handle colliers for the European coastal trade of up to some 60,000 tons.

The part played by transport costs in the total delivery costs of energy varies widely, of course, according to the form of energy and the type of transport system used. Major factors to be taken into account include the distance to be covered, the quantities actually transported, continuity of supply and average load factor and, above all, the ratio of calorific content or heat value to volume or weight. It is in these respects that economies of scale assume paramount importance. For instance, the specific capital outlay for one of the new supertankers of 200,000 tons or more deadweight is only about a quarter of that for a tanker of some 20,000 tons—or the equivalent of $70 a ton and $280 a ton respectively. When the power required to achieve a comparable speed of, say, 16 knots and the size of crew for the larger ship are taken into account, the costs per ton-km (i.e. the cost of carrying one ton of oil over a distance of one kilometre) in the larger ship works out at approximately one-third of that for the smaller vessel.

If the cheapest form of transport for conventional fuels is by large ocean-going tanker carrying bulk cargoes of crude oil, it is a remarkable fact that the cost of transporting a ton of crude oil in a 100,000 ton supertanker over a distance of 6,000 km, which works out as US cents 0·04 per ton-km, is forty times dearer than that of transporting a quantity of uranium fuel of equal calorific value by sea and rail over a distance of 1,000 km. But to maintain a correct sense of proportion, it should perhaps be recalled that the capital costs of nuclear plants are very much higher than those of conventionally fired plants and that fuel costs account for no more than a third of total costs. A more purposeful comparison is that between oil and coal transport costs and here, as Table 4 clearly shows, oil has a significant advantage.

Let us look for a moment at some of the technical and financial problems involved in the development and marketing of one of today's so-called 'new image' fuels. In the case of natural gas, the real breakthrough in transportation came as a result of the successful commercial application on a large scale of the gas liquefying process whereby it is possible, and practicable, to reduce the gas to one-sixhundredth of its original volume by lowering the temperature to about −285 °F. It is instructive to look at this for a moment in greater detail, for the developments of the past two years in the field of natural gas provide a particularly apt illustration of the technological revolution that is going on about us in energy transportation. Liquefied natural gas is carried from Arzew in Algeria to Canvey Island in the Thames estuary and to Le Havre on the French Channel coast in specially constructed tankers. Two vessels were designed and constructed for this purpose by a firm called CONCH International Methane Ltd and were baptized respectively *Methane Princess* and

Methane Progress. These vessels are identical: both have a double hull construction around the cargo space which, for greater safety, has been divided into three insulated holds separated by coffer dams. Each hold is then further divided into three aluminium tanks. The French have also built a specially designed ship, the *Jules Verne*, to bring regular shipments of natural gas from Arzew to Le Havre. On arrival at the Gas Council's terminal at Canvey Island the liquid natural gas is discharged into big aluminium storage tanks. From there it is later pumped to evaporators in which it is converted to gas by means of heat exchanged with sea water. Once re-converted, the gas is ready to flow into the national gas grid.

North Sea gas arrives in this country through pipelines. The first arrivals of gas from the British part of the North Sea occurred in 1967 and reached the coast at Easington in Yorkshire and Bacton in Norfolk through 16 in. submarine pipelines laid along the sea bed. The procedure is for the pipe to be laid by sections, each about 40 ft long, coated in bitumen and concrete and laid from a barge. Once on board the barge the sections of pipe are welded together, connected at the joints, and laid along a 4 ft trench in the sea-bed.

Overland pipelines are today a commonplace and universal tool of the big oil companies, but submarine pipelines are still comparatively new (although they have been used to a considerable degree in the Middle East) and expensive. The cost of submarine pipelines varies very substantially according to the depth of water they traverse. Recent reports quote figures of £114,000 per mile for a pipeline designed to carry 500 million ft^3 a day in shallow water, and up to £220,000 in 250 ft of water. It has been calculated that a 207 mile-long oil or gas pipeline designed to transport 200,000 barrels a day could cost as much as £95 million, or £460,000 a mile. The high figure is explained by the fact that this would be from some of the deepest waters of the North Sea, i.e. well over 200 ft in depth, where the only means immediately available would be to lay multiple small-diameter lines of, say, 8 in. This would of course have a serious effect on overall costs. When looking at these figures, however, it should be borne in mind that they relate to pipelines from the deepest parts of the North Sea and that only a relatively small proportion of the total amount of gas discovered in the North Sea is likely to come from such areas. Nevertheless, even working on a lower range of figures, it is clear that the laying of submarine piplines is an expensive operation.

It is modern science that has provided the oil companies with the new and effective methods they needed to locate potential oil and gas-bearing

Table 4. *Some relationships between average costs of energy transfer per ton-km by different means of transport*
(Calculations refer to one ton-km of coal equivalent)

Form of energy and mode of transport 1	Main transport specifications 2	Characteristic transport cost (US cents per metric ton-km) 3	Average cost index (cheapest method for each form of energy shewn = 1·0) 4
Crude petroleum			
Ocean tanker	80,000 tons deadweight—6,000 km	0·04	1·0
Pipeline	24–30 in. and 5–15 × 10⁶ tons/year	0·10–0·15	2·5–3·8
Petroleum Products			
Coastal tanker	5,000 tons—over 1,000 km	0·20	1·0
Inland waterway	600 tons—500 km	0·40	2·0
Pipeline	1 × 10⁶ tons/year—1,000 km	0·50	2·5
Rail	40 × 10³ tons cars—500 km	0·70	3·5
Road	30 × 10³ tons or under—500 km or under	1·50	7·5
Hard coal			
Ocean collier	Medium size—6,000 km	0·08	1·0
Inland waterway	500 km	0·28	3·5
Shuttle train	500 km	0·30	3·7
Pipeline	200 km	0·35	4·4
Normal rail	*c.* 300 km	0·75	9·4
Gaseous fuels			
Ocean tanker	Medium size—7,500 km	0·20	1·0
Pipeline	1,000 km	0·30	1·5
Electricity			
High-voltage transmission	500–220 kV—*c.* 500 km	0·70–2·10	1·0–3·0
Nuclear fuel			
Ship and rail	*c.* 1,000 km	0·01	1·0

Source: United Nations E.C.E. study of 'Movements of Energy in Europe and their Prospects', October 1968.

submarine strata. Of these probably the most widely known is the seismic reflection method.[1] This is really rather like an echo-sounding device, although the sound is in fact an explosion which produces sound-waves that penetrate the sea-bed and reflect back from the hard layers of rock

[1] Originally developed for use on land, it has in recent years been used increasingly for marine surveying.

that lie beneath it. The usual practice is for boats to sail over carefully selected and pinpointed stretches of water and, by means of the detonation of a series of explosions, to enable the geologists on board to prepare a map showing the shape of the rock layers beneath the sea. But although these seismic measurements narrow down the field and give invaluable indications as to where oil and gas might be found, they do not automatically prove or demonstrate their presence. This can only be done by drilling.

Drilling at sea is popularly associated with giant platforms towering high above the surface of the water in the midst of swirling mists. For those on board, however, it is a normally undramatic life full of hard work and isolated from the rest of the world. That it can also be a hazardous occupation was brought home only too cogently by the tragic loss of the B.P. rig *Sea Gem* on 27 December 1965.

As in the case of Saharan natural gas delivered to the Canvey Island terminal, storage areas or tanks will have to be provided for North Sea gas. A certain amount can of course be carried in the national gas grid, but this amount is marginal in relation to the total quantities involved. The position is complicated by the fact that there are very large variations in the levels of demand for gas (as, indeed, there are for most other fuels, notably electricity), not only over the year and season, but also over any given day. It is interesting to note in this respect that, in the United States, as a result of the widespread use of air-conditioning, the demand for electricity (the main form of energy used) is far less subject to seasonal fluctuations than in Western European countries. Short-term variation in demand can usually be met relatively easily by drawing upon gas reservoirs in the conventional and well-known, if little loved, gasholders. These are normally capable of holding between $2\frac{1}{2}$ and 5 million ft^3 of gas, although there are a few with a capacity of up to 10 million ft^3. But gasholders are not the answer to major seasonal, or even daily, fluctuations, and will be even less able to cope as the demand for gas grows over the next few years. According to the information services of the Gas Council the gas industry requires at the present time something like 70–80 thousand extra ft^3 of gas during the winter demand period. To meet this would require the construction of many thousands of additional gasholders, which would be economically ruinous and aesthetically unacceptable. The answer so far has lain in utilizing a certain number of plants, usually the most modern and efficient, as baseload plants and bringing in other plants, normally the oldest and least economic, during peak-load periods. This of course inevitably means that a number of plants pressed into service during the winter stand idle during the summer months. Clearly, from the economic point of view it would be much more advantageous if gas could be continuously produced from baseload plant, and any surpluses developed during the summer months stored in suitable reservoirs. The same argument applies even more forcibly to natural gas from the North Sea, both in order to build up reserves against peak demand periods and, above all, as an insurance against any interruption in supplies as a result of technical difficulties or mishaps.

A great deal of work has already been carried out in the United States, as well as in a number of European countries, on the storage of natural gas. In the United States, for example, natural gas from the gasfields of Texas and the Gulf is piped to the industrial areas of the north-east and is stored in the old and exhausted gasfields of the Appalachian Mountains. In the Soviet Union the Russians are using old saltmines as underground reservoirs for gas storage purposes and are building a series of vast underground reservoirs around the Moscow district. In Italy plans are in hand to store imported natural gas from Libya or elsewhere in exhausted gasfields in the Po Valley. Finally, the French have adapted underground caves near Lussagnet as a gas reservoir.

In the United Kingdom the Gas Council has been looking for a number of years for suitable sites for storing gas underground. Their search is based on the geological evidence that throughout the world there are a number of natural underground 'domes' created as a result of lateral rock pressures. Provided one of the upper layers of the 'dome' consists of impermeable rock, gas can be pumped down through wells until it reaches suitable water-bearing strata (sometimes known as aquifers). The gas, being lighter than the water content in the pores of the rock, will rise above it and so form an artificial gasfield. An example quoted by the Gas Council Information Services referred to an elliptical 'dome' with axes about four miles and two miles long. A structure of this kind would provide storage capacity for some 15,000 million ft^3 of gas.

But while underground storage in aquifers would provide a satisfactory solution to the problem of meeting major variations in seasonal demand, it is less satisfactory for dealing with sudden sharp variations. Since, however, the quantities involved in the latter case are not so great, it is possible to provide for this situation by storing natural gas as a liquid in specially constructed tanks. In essence, this method consists of pre-freezing an earthen cylinder of the required diameter and depth, excavating the unfrozen earth from the inside and covering the cavity with a gastight and suitably insulated roof. The walls and floor of the earthen tank are kept frozen by the cold liquid it contains. The liquid is withdrawn as and when required and can be easily re-gasified. Several such tanks, 130 ft in diameter and 130 ft deep, capable of holding 21,000 tons of liquid natural gas (equivalent to almost 11 million therms) have been constructed by the Gas Council at their Canvey Island terminal.

With the gradual emergence in the last ten years or so of a commercial nuclear power industry, yet another blow is being struck at what was, until comparatively recently, the veritable stranglehold exercised by the geographical absence or lack of fuel resources. Today, one ton of unenriched uranium[1] can provide as much heat as 8,000 tons of coal. At the same time, it is, of course, infinitely easier to transport. Although nuclear power stations, because of their high initial capital costs, must necessarily run as baseload stations—thus effectively limiting their construction at the present time to areas that already have a considerable degree of industrialization—they are virtually independent of transport cost considerations. On the other hand they also require, in addition to a high initial capital investment, highly skilled and scientifically trained personnel and, on the operational side, vast quantities of water for cooling purposes. But the economics of nuclear versus conventional fuels costing exercises is a highly complex matter, subject to rapid swings and changes in the cost pendulum, to which we shall return later.

The net result of these developments in the transport field has been a dramatic change in the energy supply situation. Whereas, for example, in the late 1950s there were substantial variations in the price of energy (whether coal or oil) in different parts of, say, the Federal Republic of Germany, by the middle of the 1960s these differences had narrowed considerably and oil industry salesmen were confidently predicting the imminent equalization of cheap oil prices over the entire country as a result of the construction of refineries at the major inland consumer centres. In this respect, Europe has once again followed where the Americans had long since led the way.

All these factors have fundamentally altered the overall pattern of energy consumption—although not preventing and, indeed, greatly accelerating—the switch from one form of energy to another. With the ready availability of oil and natural gas in all parts of Europe, coal, which once ruled supreme, has suffered heavily. With the best seams long since exhausted, with an industrial structure that is still basically labour-intensive, and obliged to operate in geological conditions that are often difficult, the

[1] One ton of enriched uranium is usually calculated as equal, in terms of heat generated, to 50,000 tons of coal.

coal industry of Western Europe has inevitably been burdened with higher production costs. In the United States, for example, where coal mining has become to all intents and purposes a capital-intensive industry, and where geological conditions are many times easier than in Europe, output per manshift in 1966 averaged some 16 tons, compared with 2·3 tons at the best European (i.e. German) mines (and a figure of 1·8 tons in the United Kingdom). Coal, with its essentially local or regional markets, is, at the same time, infinitely less flexible in its marketing potential than its chief rival: oil. The big oil companies, with their wide international operations and ramifications may to some extent, be compensated for lower prices in one market by higher prices in another. Thus, as we have already had occasion to see, the United States has, despite its alleged support for the principles of freer trade, adopted an energy policy which is a great deal less liberal than that of a number of European countries and designed to ensure a rate of return on indigenous oil production that guarantees the maintenance of output at a strategically acceptable level.

This, in turn, has had three consequences: first, since indigenous United States produced oil is substantially dearer than Middle East oil (even after allowing for cost of transportation), it means that any imports that are allowed into the country are sold at prices aligned, in an upward direction, on the price of American oil, thus fetching handsome profits; secondly, if oil imports into the United States were freed, much more Middle East oil would flow there, thus releasing the pressure on European markets and causing a hardening of oil prices; thirdly—and perhaps most importantly of all—it has provided the American oil companies with a secure and lucrative home base.

Before studying all the social, regional and strategical considerations, however, we must first take a closer look, in Section II, at the importance and effect of energy prices and availability upon industrial costs and industrial policy.

SECTION II

INDUSTRY AND THE PRICE OF ENERGY

What, it may be asked, are the reasons that make industrialists choose one area in preference to another for a particular factory or plant location? From the mass of papers and documents that have been written on this subject it seems that, while there may be many, sometimes mutually conflicting, and sometimes temporary, reasons, three main factors stand out:

(i) access to markets;
(ii) access to labour; and
(iii) access to raw materials.

For the industrialist energy is a raw material. Its relative importance both in the raw material category and in the overall costs spectrum is, of course, a factor of the amount of energy required in the production process. In cigarette manufacturing, where energy costs are slight, energy represents an almost negligible proportion of total costs. To a steel or aluminium plant, on the other hand, energy costs are decisive.

The importance of the energy cost factor

The importance of energy costs in relation to total costs varies from country to country and from industry to industry. The 1966 Community study,[1] to which we shall be referring very frequently in this Section, found that in ore and mineral mining energy costs ranged from 5 per cent in France to 15·3 per cent in Germany; in the chemical industry, from 7·1 per cent in France to 13·8 per cent in Holland; in steel, from 12·6 per cent in Holland to 21 per cent in France; in non-ferrous metals, from 2·3 per cent in Holland to 10·3 per cent in Germany; and in transport, from 5·3 per cent in Holland to 12·9 per cent in Italy. Some of these differences are, however, accounted for by differences in statistical definitions and in the structure of certain industrial groupings, as well as by differences in energy prices and levels of specific consumption.

Table 5 shows wages, amortization and energy costs for six selected

[1] L'Influence économique du prix de l'énergie, *op. cit.*

industries in France and Germany. The differences between the two countries are, on the whole, of a comparatively minor nature and may be largely attributed to reasons of a structural character. Variations within the energy cost sector, with the exception of the chemical industry, are relatively unimportant.

Table 5. *Salary, amortization and energy costs in selected industries in France and Germany*

	Salary and wages		Amortization		Energy costs	
	France	Germany	France	Germany	France	Germany
Electricity generating	20·4	13·1	17·5	13·3	20·1	24·9
Steel	15·9	16·9	9·1	5·7	21·0	20·9
Chemicals	22·9	20·1	4·8	5·9	7·1	12·2
Textiles	21·5	27·9	3·2	5·1	2·8	3·7
Non-ferrous metals	16·2	15·5	4·9	4·2	8·0	10·3
Shipbuilding	23·0	21·9	4·2	4·0	3·0	3·1

The Community Study found that the industries for which energy costs represented more than 10 per cent of the value of production accounted for nearly 15 per cent of the total industrial output for the Common Market; those for which energy costs represented 5 per cent or more of the value of their production, contributed 23 per cent to total industrial output. There were, inevitably, some considerable differences from one country to another. Taking first those sectors where energy costs represented 10 per cent or more of the value of production, it was found that they accounted for 17 per cent of total industrial output in Belgium, 16 per cent in France and Germany, and 14 per cent in the Netherlands. When a figure of 5 per cent was taken for purposes of comparison, the figure for Belgium leapt up to 36 per cent while France, Germany, Italy and the Netherlands were all grouped closely together with a range of 21 to 23 per cent. Another interesting feature to emerge was the fact that, although salaries and wages represented, as might have been expected, a higher cost element than energy in almost all cases, energy costs were, in their turn, more significant than those arising from amortization over a very wide range of industry.

It is, however, not only the direct costs of energy for given industrial consumers that determines the full extent of the influences of variations in fuel prices. Of equal, or indeed in many cases of considerably greater, even if more indirect, influence, is the effect increases in energy prices may have upon other energy intensive products required by the industries in question. These so-called secondary results have to be added to the immediate or primary effects of a change in energy prices if the latter's full effects are to be accurately calculated.

The Community study paid particular attention to the importance of energy costs for exporting industries. It was found that direct energy costs played a much more important role in the production of goods earmarked for export than in those destined for general internal consumption. When the value of energy exports was added to the direct energy costs incurred in the manufacture of industrial goods sold on the export market, a figure of about 10 per cent was obtained. Even when real energy exports were omitted from the calculation, energy costs still represented a proportion of 4 to 6 per cent, according to country, of the total costs of export goods, i.e. marginally above the figure for direct energy costs for total industrial output. This was explained by the fact that the more energy intensive industries are generally to be found among the main exporting industries. Thus, industries for which energy costs represent 10 per cent or more of the total value of their production accounted for 22 per cent of all exports, while those for which energy costs represented more than 5 per cent of the value of their production contributed 39 per cent. In individual countries the figures, on the 10 per cent basis, were 20 per cent for Belgium, 30 per cent for France, 24 per cent for Germany, 32 per cent for Italy and 21 per cent for the Netherlands; while the corresponding figures on the 5 per cent basis were 54 per cent in Belgium, 43 per cent in France, 36 per cent in Germany, 33 per cent in Italy and 40 per cent in the Netherlands.

Energy prices and industrial competition

The ability of an industry or enterprise to compete in world markets depends primarily on the prices it is able, or obliged, to charge for its products. These prices, in turn, depend upon the cost of the various elements that make up a total industry's production costs, to which there must normally be added a reasonable margin of profit. The cost to the industry of these various elements may rise or fall, though usually the former, and the industry can seek to respond by raising its prices for its own products, by absorbing increased costs through the acceptance of lower profits, or by reducing its purchases or intake of raw materials. Normally, the first or second of these alternatives is preferred, since the third would presumably result in a lower level of output for the industry's products. In order to try and assess the effect of an increase in energy costs, the Community study took as a working hypothesis the assumption that only energy prices had been changed and that all the other cost factors had remained unaltered. This was, admittedly, an artificial concept but had the merit of enabling the effect of energy price increases to be computed, at least theoretically, with greater precision. Any change in price must of course affect the flow of trade and thus modify in some way the cost factors of all the industries or enterprises involved. The second assumption made in the Community study was that individual enterprises or industries would, in fact, effectively pass on any increases in energy costs by raising the prices for their own products, thus triggering off a chain reaction. By following this process through to the final consumer stage, the study attempted to calculate the cumulative effects of an energy price increase. The Community study also assumed that the energy price increase was a once-and-for-all affair, i.e. that the resultant price increases in other sectors of the economy would not in turn trigger off further energy price increases and that indirect taxes would increase pro rata with prices. This was admittedly an imperfect assumption but was regarded as paving the way for a second and later more detailed calculation of the effect of energy price increases based on the premise that entrepreneurs and business management tended to work on the basis of a fixed relationship between profits and turnover.

The extent of the initial increase in energy costs was not considered to be of fundamental importance—the more so since it was assumed that it would in any case affect only the prices of other products and not the actual volume of trade. As a result the connection between the price of energy and the price of industrial products was taken as a linear relationship which remained unaffected by the actual amount of the increase in energy prices. Assuming therefore that with a given product *Y* the price of energy supplies was increased by *X* per cent the result would be an increase in the price of *Y* of $\left(\frac{X.Y}{100}\right)$ per cent.

Effect of an increase in the price of energy on different industries

It was found that in all six Member countries the biggest variations in end-prices as a result of higher energy costs occurred in some four industries: steel, chemicals, non-ferrous metals and paper and board.

In the steel industry, the biggest increases occurred in France and Germany, while the effects were considerably less in Italy and in the Netherlands. These variations were explained by differences in production techniques. In the chemical industry, on the other hand, the biggest increases were found in the Netherlands (where the chemical industry was, still at that time, largely based upon coal) and the smallest in France. In the case of non-ferrous metals, glass, pottery and paper there were no significant variations among the six countries. Changes in the price of natural gas, it was stated, had a considerably greater impact than changes in the price of coal or oil.

The immediate or primary effects of energy price increases were found to be of particular importance for raw-materials industries and of comparatively minor importance for the majority of processing industries. Thus, in the steel industry, for example, the primary effect of an energy price increase is more than twice as much as that of the secondary effects. In the clothing industry, on the other hand, secondary effects were found to be some ten times bigger than the primary effects. Other industries range between these two extremes.

An attempt was also made to try and assess the comparative importance to industry generally of price increases for energy, and chemical and steel products. This suggested that there were only a handful of industries where increases in the price of steel or chemicals were more important than an energy price rise. In the case of steel, these were the car and shipbuilding industries; for chemicals, the rubber, asbestos, textile, leather and clothing industries. In all other cases the effects of an energy price increase were regarded as being more significant.

In conclusion, this part of the Community study looked at the effect of energy price increases on the price levels of the different component parts that go to make up the global demand figure. The global demand was divided for this purpose into consumption, investments, exports and stock variations. Clearly, the effect of energy price changes was not identical for all four categories. The area that was most severely affected was exports, where the increase was of the order of 20 per cent. For the rest, the effects were felt in descending order by private consumption, basic industries and investments.

Effect of the price of energy on the location of industry

It is axiomatic that a country's economic expansion depends upon continuing industrial development. Where, within a given country, there are major disparities in prices for raw materials or labour between regions, this immediately creates distortions which can have a severely harmful effect upon sound economic progress. As far as energy is concerned, the whole scene has been dramatically altered by the changed relationship of the various fuels, and in particular of coal and oil.

As far as the producers of energy are concerned, location must inevitably remain largely a question of where the physical resources happen to be found; this is certainly the case for coal, oil and natural gas, though in the case of the latter two fuels the development of the pipeline and the giant tanker has enormously increased the flexibility of refinery location. The refineries themselves are increasingly located on or very close to major areas of consumption. An opposite trend is perceptible in the construction of electricity power stations where the cost of fuel transportation, for conventional plants at least, is usually considerably greater than that of high-voltage electricity current. This pattern may change rapidly, however, at least in the medium term, with the emergence of nuclear power. Nuclear stations burning ordinary non-enriched uranium, one ton of which, it will be recalled, provides as much power as 8,000 tons of coal, are largely independent of fuel transport considerations. Their high initial capital cost, on the other hand, makes them dependent on base-load operation and this in turn encourages their construction as near to major consumer areas as safety and health considerations will permit. For the steel industry the location of its basic raw materials, i.e. coking coal or coke and iron ore, is still of paramount importance since their transport costs are high—and considerably greater than those for finished and semi-finished steel products.

If for the producers of energy, access to raw materials remains a determining factor, consumer location preference is subject mainly to market proximity and manpower availability. Clearly, however, no assessment can be made without knowing exactly those industries for which energy costs are a determining, or major, factor in the choice of location. In some countries, of course, there are only minor differences from one area to another. In these cases other factors will determine the issue. In fact, energy costs normally have an important bearing on choice of location when they exceed 5 per cent of total costs. In cases where energy costs represent less than 5 per cent of total costs their influence becomes so slight that even big regional fluctuations do not affect the fundamental choice. Clearly, however, even for industries for which energy costs represent more than 5 per cent of total cost it is far from being the sole or even major determining factor—the final choice of location normally being that combination of all the factors concerned which is considered to be the most favourable from the firm's point of view. By way of illustration four industries, the aluminium industry, the cement industry and the steel and chemical industries, were examined in closer detail.

Aluminium industry

The main considerations underlying the choice of plant location in the aluminium industry were found to be energy availability—since energy costs account for approximately 30 per cent of total costs in this industry—together with the cost of transport of raw materials, principally bauxite, and of finished products to consumer centres. Supplies of cheap electricity were described as a *sine qua non* for the entire operation: hence the traditional aluminium industries of countries like Norway and Canada with large and cheap hydro-electric resources. Cheap electric power had similarly determined the location of the bulk of the German aluminium industry. The whole discussion of the development of an aluminium industry in the United Kingdom is dependent upon the availability of cheap power, whether from a coal or nuclear origin. The difficulty in this case arises from the fact that the Central Electricity Generating Board (C.E.G.B.) consider that they are, under their statute, obliged to desist from price discrimination. They have maintained, in other words, that they are required to charge similar prices for comparable consumers and, *ipso facto*, to refuse to earmark part of the output of electric power from a nuclear station—at an exceptionally favourable price—for an aluminium smelter plant. The strict application of this thesis would of course make the aluminium projects in the United Kingdom financially unworkable. The way out has been for the aluminium companies to generate their own power and this is, for example, what Alcan will be doing in Northumberland. This has understandably led to protests from the C.E.G.B. that they have to pay an average price for their coal supplies over the country as a whole of over 5*d.* a therm, whereas Alcan will be getting its coal supplies at a price alleged to be in the neighbourhood of 3*d.* a therm.

Cement works

This is an industry in which energy is used both as a source of heat and as a raw material. The share of energy costs in total costs is put at about 20 per cent, two-thirds of which are represented by primary fuel requirements and the balance by electric current. The choice of plant location is regarded as being subject to the ready supply of raw materials, i.e. the proximity of adequate supplies of suitable types of earth for cement-making, blast furnace slag for metallurgical cement, and energy in the shape of coal, oil and electricity. Of all these, the availability and above all the proximity of suitable types of earths is undoubtedly the most important single consideration. It represents by far the largest and most bulky factor in quantitative terms and loses much of its weight in the process of transformation. The relatively plentiful resources of suitable clays and other earths have had the effect of reinforcing and inflating the relative importance of energy prices.

Steel

The steel industry is the biggest consumer of energy outside the electricity generating industry, though it is important to distinguish between the respective requirements of the different elements that go to make up what is broadly described as the steel industry: i.e. blast furnaces, steelworks, rolling-mills, foundries and iron works. Although the choice of location is often simplified by the fact that many plants are completely integrated, the basic objective is to find the most suitable conditions for the production of pig-iron.

The main blast furnace requirements are for adequate and regular supplies of coke and iron ore on the one hand, and easily available outlets on the other. In fact sites are normally selected to reduce transport prices for both coke supplies and iron ore to a minimum, even at the cost of constructing at greater distances from those markets taking the crude steel output. The decision whether to build at a site close to coke or coking coal supplies, or, alternatively, to iron ore fields, is governed by the weight of relative consumption of both these essential raw materials. There are in Western Europe, for example, few economical or commercially attractive orefields (with the exception of Sweden) with the result that the majority of the more traditional European steel works are concentrated in or near to the coalfields. In recent years, attracted by the prospect of imports of cheap American or Polish coking coal, there has been a move towards the construction of new steelworks near the coast, i.e. Dunkirk, Zelzate (near Ghent in Belgium) and Taranto—as well, of course, as the recent proposals for the construction of a new big steel complex on the Clyde in Scotland. With European iron ore mines nearing exhaustion, and with ever increasing imports from abroad (i.e. from Africa and Australia) the advantages of coastal locations are still further enhanced. The drawbacks are, first, the greater distance, in many cases, from the markets for finished steel products and, secondly, the possibility that American coking coal supplies may, in the longer term, not provide the security of supply that was, until quite recently, expected of them. The importance attached to the conservation of coking coal production in the Community was demonstrated by the decision taken in 1967 by the Governments of all six Common Market countries to grant a subsidy of $1·7 a ton (with provision for raising this to $2·2 a ton in certain cases) of all supplies of indigenous coking coal supplies to Community steelworks. The total amount involved in a full year (i.e. 1967) was of the order of $64·4 million.

A great deal of research and development work is currently in progress in Germany, Japan and the United States, on the direct reduction of iron ore in steelmaking by means of the elimination of blast furnaces and, consequently, the need for coal. In this connection it was stated by Mr Finniston, Deputy Chairman of the British Steel Corporation, in an address delivered to the International Steel Congress held in Los Angeles in November 1968, that recent studies had shown that, in Britain, with its comparatively high energy prices, steel could now be produced with a direct reduction/electric arc process (i. e. in other words with no need for coke) at a cost that was only some 20*s*. a ton higher than in the large modern blast furnaces using oxygen injection and requiring coke in steelmaking. If, he continued, British steelworks were able to obtain natural gas or electricity supplies at half the present price, as was possible in certain other countries, then the direct reduction process used in conjunction with an electric arc furnace would be competitive with blast furnaces using oxygen injection techniques. From this Mr Finniston concluded that the next ten years could reasonably be expected to witness a considerable development of the direct reduction process, particularly in those countries with cheap indigenous energy resources or a cheap energy import policy.

The part of fuel costs in total costs for an integrated steelworks has been estimated at approximately 20 per cent: while their impact upon the various stages in the steelmaking process is estimated to be broadly as follows:

Table 6. *Breakdown of energy costs at various stages of production in an integrated steelworks*

Steelworks	Energy consumption (%)
Blast furnaces	66
Steelworks	12
Rolling-mills	19
Others	3
	100

The process of integrating blast furnaces and iron works began towards the middle of the nineteenth century when the latter began to search for places where they could get, at one and the same time, secure supplies of pig-iron and markets for their products. Moreover, the rapid development of the continuous cycle embracing the manufacture of pig-iron, steelmaking and rolling, enabled big savings in fuel consumption to be made. The excess gases produced by the blast furnace operations which were wastefully flared off in the early industrial period were another economic factor making for integration of the various manufacturing processes.

The question has been raised—quite separately and distinctly from the potential development of the direct reduction process—to what extent the continuing fall in the specific consumption of coke—from the 3·5 tons of coke required for each ton of steel in 1850 to 1·5 tons in 1913, 0·950 tons in 1964 and a possible figure of 0·400 to 0·500 by the early 1970s—may affect the location of existing steelworks in the traditional coal mining areas. Many experts agree that a new trend is already clearly discernible: i.e. the construction of blast furnaces and steelworks near to the coast, as at Dunkirk, to process the ores and coke, followed by the transport of the crude steel to rolling-mills situated near to the points of consumption such as ship-building yards and car factories. This is, notably, the policy that has been followed by Sidmar, the big new steel plant built by a consortium of Belgium, French and Luxembourg steel companies near Ghent in Belgium, as well as by a combine of 15 German enterprises and the Dutch steel company, Hoogovens, near Rotterdam. In both these cases, iron, steel and semi-finished products will be produced at the new coastal plant and then transported to inland plants in the Ruhr, the Liège basin and the Luxembourg and Lorraine producing areas for further processing.

Chemical industry

Germany was quoted as providing a typical example of the origins of one of the world's leading chemical firms, with a history going back to the second half of the nineteenth century. This particular company, IG Farben, began its existence along the confluence of the Rhine and Main rivers. Although regarded originally as useful transport avenues these rivers soon became important in their own right as sources to meet the industry's growing needs for water.

Originally, energy was not a major factor in the choice of location for the chemical industry. This phase, however, was short-lived and the growing use of coal laid the foundations of the carbo-chemical industry as, for example, with the Dutch State Mines in Holland, and big coal, steel and/or chemical works complexes in Germany and, to some extent, in the United States. Similarly, in more recent times, oil refineries have played a major part in determining the location of new chemical works, although perhaps the most impressive example of all in postwar Europe, at least up to the present time, has been the creation of a big chemical complex in the Lacq area in south-west France, following the strikes of natural gas there in the early 1950s. Recently, of course, carbo-chemicals have been giving way to petro-chemicals, and this has liberated the chemical industry from any kind of locational dependence upon coal. Since, moreover, the price of oil tends today to vary comparatively little from one area to another in any given country—at least in Western Europe—the cost of energy has lost a good deal of its original importance. Nevertheless, despite its undoubtedly diminishing impact, the energy price factor cannot be disregarded. It would appear, moreover, that whenever a big chemical firm wishes to expand into a new area, it is usually able to persuade one of the big oil companies of the advantages of joint or parallel expansion. Indeed, the increasing partnership between the oil and chemical industries has become a major feature of the modern industrial scene.

Community experts drew the following conclusions from the examination of the effect of energy prices on the location of industry:

(i) That energy costs could not be regarded as constituting a dominant influence on the choice of industrial location of more than a very limited number of enterprises. Thus, even in those industries where the proportion of energy costs to total costs is high, these had normally only a complementary as opposed to a decisive bearing on the final decision.

(ii) The impact of energy costs was, if anything, likely to diminish rather than to increase in future years: the specific consumption of energy in most industries was tending to decline, and regional variations in energy prices, which until recently were considerable and general, were also falling against a general background of more plentiful energy supplies and greatly enhanced substitution possibilities. Moreover, even for the big energy-consuming or energy-intensive industries, the differences in transport costs for other raw materials as well as for their finished products remained of very great importance, though exceedingly difficult to translate into comparative financial terms. These differences are far less likely to diminish in the future than those in energy prices (whether due to the use of different fuels or for transport reasons) and can be expected therefore to grow in relative importance. Manpower availability is another factor of growing importance, particularly in countries, like Germany, that are already confronted with acute shortages in this respect. Other factors, such as the price of land, building materials and industrial rents are also acquiring greater relative importance in these circumstances.

(iii) Nevertheless, although for the economy as a whole, the importance of energy costs as a prime factor in determining the location of industry seems likely to diminish rather than increase, it remains one of high significance and particularly so for the economically or industrially less favoured areas. Bearing in mind the question of the further potential reductions in energy costs arising from its more efficient use, and the development of new sources of energy and new techniques to exploit them, it is undoubtedly a sector where firms could usefully seek to obtain further cost savings. Indeed, the prospects for securing further cost reductions in firms' expenditure on their energy requirements seem considerably brighter and much more within the compass of their own freedom of action than in the related fields of raw materials and manpower.

The influence of the price of energy on regional development

There are wide and deep-rooted differences in regional structure and regional rates of development in nearly all advanced industrial economies. It is perhaps self-evident, but nonetheless true, that the success or failure of regional policy depends upon the success or failure of the industries in that region. It depends as much on the renovation of established industries, for example by branching out into new fields, as on the ability to attract new industries. The post-war history of areas like Northern Ireland or the north-east of England has demonstrated the dangers of regional dependence upon a small number of traditional industries. Quite apart from the social, economic or industrial difficulties which such a situation may give rise to, the resulting problems have been compounded by the widely held view that such regions, often scarred by an ugly industrial heritage, are cultural and aesthetic wildernesses, thus discouraging many university graduates or men of similar calibre from considering the possibility of making their careers in them. When matters have reached such an extreme situation, the only solution is a radical root-and-branch approach which couples the establishment of new industries, often with state assistance, with urban renewal and the creation, and if necessary the subsidization, of adequate facilities. Inevitably, the cost of such operations is high, particularly in a country like England with its many appallingly hideous and soulless industrial conurbations. But if the price is high, it is also the only way to turn these slums of the industrial revolution into decent, worthwhile and tolerable centres of communal existence.

Attempts have been made in the European Community to measure the effects of energy costs on the regional policy of a number of key industries. A distinction was drawn between the essentially regional, or localized, industries, for which transport costs are often a determining factor for the price of their end-products and for which their geographical isolation often constitutes a form of protection, and industries that are national or international in character, which tend to rely only to a very minor extent on local demand, or have no geographical protection, and are consequently subject to the full force of competition. It is the firms that fall into this second category that are particularly important to a successful regional structure or policy since they have the power to attract to themselves a wide range of other industries geared to provide them with many of the factors of their production processes, i.e., a chemical plant leading to the building of an oil refinery, cracking installations, engineering works and service industries. In this way the advent of one big enterprise can generate intense activity and employment for the local and regional economy. Such key industries, and the industrial chain reaction which they set in motion, are commonly referred to as 'polarization' industries.

Economic history provides a number of examples of the importance of the price of energy in regional development. This applies with particular force to the iron and steel industry which has a remarkable record of industrial mobility through the ages. It was, for instance, lack of timber and its consequent rise in price (coal did not conquer this market till the next century) that led to the decline of iron production in eighteenth-century England; at the same time there was a big rise in Sweden where both iron ore deposits and wood from the forests for the charcoal ovens were plentiful. In the next century, the development of the blast furnace burning coke and the elimination of the old charcoal ovens had a profoundly modifying effect on the economic geography of Europe. The traditional ironworks of Wallonia, in eastern Belgium, Lorraine, and the north of France were obliged to move out of the plains to new sites near the coalfields or iron ore mines and near to water. Even so, these changes were not fundamental as the industry tended to remain within the same broad economic area. In England the abundance of coal laid the foundations of the country's industrial and military greatness, though at a cost, as Daniel Defoe bore witness as long as two centuries ago:

> such has been the bounty of nature to this otherwise frightful country that two things essential to the business, as well as to the ease of the people are found here, and that in a situation which I never saw the like of in any part of England: and, I believe, the like is not to be seen so contrived in any part of the world; I mean coals and running water upon the tops of the highest hills. This seems to have been directed by the wise hand of Providence for the very purpose which is now served by it, namely, the manufactures, which otherwise could not be carried on; neither indeed could one fifth part of the inhabitants be supported without them, for the land could not maintain them.

Other regions were less fortunate. Thus in France, areas such as Perigord, Savoy, the Jura, Champagne and Normandy—all of which had in the eighteenth century been important iron producing regions—faded from the economic map of France because of the complete impossibility of securing coal at prices that enabled them to remain competitive with their more favourably situated rivals. Despite the fact, therefore, that coal costs were not the only factor, it was this that either destroyed them altogether or reduced them to a relatively minor status. While, in more recent times, two of these regions, the Jura and Savoy, have been able to develop important mechanical construction and electro-metallurgical industries, this has been due to the development of the local hydro-electric or 'white coal' resources. None of the other once important industrial areas, deprived of readily available energy supplies, have been able to emulate them. Moreover, the decline of the old ironworks in areas like Champagne and Perigord also had adverse effects on related or associated industries, which in many cases soon followed the ironworks into extinction, and even on local agriculture. It is this phenomenon which largely explains the tremendous regional economic disparities in France and the striking concentration of French industries to the north and east of Paris. With the trend towards equalization of energy prices that is such a feature of the present decade, it is likely that this concentration of industrial resources in France will become less marked. But, at the same time, the effect of such factors as the pull of tradition, the momentum of industrial development and the existence of large ready-made markets cannot be under-estimated, and even a bold government policy of selective financial or fiscal advantages may only be able to change such a traditional industrial pattern after a good deal of time and patience. Some successes have already been recorded, notably with natural gas at Ragusa and Gela in Italy, Lacq in France and Groningen in Holland. Lacq is perhaps the most interesting among these examples as the French government has sought, with considerable success, to use the natural gas in the promotion of the industrial development of south-western France. The building and construction industry has shown a striking development and many French companies in this field have set up important branch offices in what was formerly a much neglected part of the country. The French chemical industry has made particularly big investments in the area and has been largely responsible for a near fourfold increase in the flow of trade in the space of ten years through the port of Bayonne. Between 1954 and 1959 the value added by enterprises in the region rose by an average of 9·6 per cent a year, compared with an average rate of increase of 4·6 per cent for

the country as a whole; family income increased by 6·5 per cent a year as against a national average figure of 4·6 per cent; while state receipts showed an overall increase of 62·7 per cent compared with 38·5 per cent for the country as a whole. State expenditure for the area, on the other hand, was less than for any other part of France. Seen against this background, it is tempting to speculate on the economic possibilities which natural gas from the North Sea may open up for areas like East Anglia which are comparatively backward industrially as well, of course, as for the renovation of the existing industrial structure of north-east England. Among the schemes that have been put forward, most of them of rather too fantastic a nature to be taken really seriously, was that of building a massive barrage across the Wash, with a view to creating at one and the same time a major new port as well as a new city located near a plentiful source of energy. Other projects have included the use of natural gas in order to promote the creation of a vast and entirely new industrial belt stretching from Humberside to Lowestoft.

It is worth pointing out, on the other hand, that the broadening availability of energy and the growing equalization of fuel prices do not result in any automatic process of industrialization. Thus, in Germany, the coming-on stream in 1963 of the first of the big refineries at Ingolstadt in Bavaria and a consequent reduction in the price of heavy fuel oil from DM. 120 a ton to first DM. 100 and later DM. 70 a ton have not led to the establishment of any big new industrial enterprises. Up to the present time, at least, manpower difficulties and the extent of the distance between local production centres and important consumer markets have mitigated against a big regional industrial upsurge. Nevertheless, the Bavarian authorities have remained confident that in the medium term the attraction of the Ingolstadt refineries will significantly encourage substantial industrial development, all the more since Dutch natural gas will soon also become available as a result of the extension of the pipeline network throughout southern Germany.

The changing pattern of energy demand

Since the end of the Second World War there has been a dramatic change in the pattern of energy. Coal, which right up until 1938—and for the first three or four years after the war—had dominated the world energy scene, has been toppled from its once pre-eminent position by the successive challenges from oil, from natural gas and, within the last two years or so, from nuclear power. As a result, while the demand for new sources of energy has soared, the absolute level of coal consumption has shown only a modest increase, as in the United States, or declined sharply, as in Western Europe. The rate of substitution has, inevitably, varied from country to country, according to the availability of the alternative sources of supply, their costs and the cost of coal production. Coal production costs are, for instance, very much cheaper in the United States, where coal for electricity power stations can be sold, profitably, for about $3·50 to $4·00 a ton at the pithead, than in Europe, where the average price is nearer £4 10*s*. or nearly $11 a ton. In other countries, like Poland, actual costs are regarded as taking second place to the need to export and so secure foreign currency. Figures showing the development of the absolute and relative levels of consumption of the principal forms of energy in North America and in the main O.E.C.D. areas and in the Soviet Union and East European countries are given below (Table 7).

As these figures show, the absolute totals for solid fuels have often continued to increase, but at a fractional rate of that for other fuels.

Table 7.

Primary energy consumption in North America: 1950–64

	Solid fuels	Natural gas	Oil	Nuclear	Hydro	Total
		(In million tons coal equivalent)				
1950	488·4	233·3	474·0	—	25·3	1,221·0
1955	435·3	352·4	622·0	—	32·0	1,441·0
1960	380·0	491·5	737·0	0·8	42·8	1,665·8
1964	426·5	612·4	828·0	1·7	48·6	1,917·2
		(In percentages)				
1950	40·0	19·1	38·8	—	2·1	100·0
1955	30·2	24·4	43·2	—	2·2	100·0
1960	23·2	29·9	44·4	0	2·5	100·0
1964	22·3	32·0	43·2	0	2·5	100·0

Primary energy consumption in O.E.C.D. Europe: 1950–64

	Solid fuels	Natural gas	Oil	Nuclear	Hydro	Total
		(In million tons coal equivalent)				
1950	494·6	1·8	85·7	—	17·5	599·6
1955	563·0	7·0	161·8	—	24·8	756·6
1960	532·7	16·0	281·3	1·1	36·3	867·4
1964	527·0	23·7	477·8	4·8	40·2	1,073·5
		(In percentages)				
1950	82·5	0·3	14·3	—	2·9	100·0
1955	74·4	1·9	21·4	—	3·3	100·0
1960	61·5	1·8	32·4	0·1	4·2	100·0
1964	49·1	2·2	44·5	0·4	3·8	100·0

Table 7 (*continued*)

Primary energy consumption in Japan: 1950–64

	Solid fuels	Natural gas	Oil	Nuclear	Hydro	Total
		(In million tons coal equivalent)				
1950	40·2	—	3·3	—	6·3	49·8
1955	49·5	0·3	13·8	—	8·3	71·5
1960	67·5	1·3	44·9	—	9·6	123·7
1964	71·6	2·9	106·4	—	11·3	192·2
		(In percentages)				
1950	80·7	—	6·6	—	12·7	100·0
1955	68·9	0·4	19·2	—	11·5	100·0
1960	54·7	1·1	36·4	—	7·8	100·0
1964	37·2	1·5	55·4	—	5·9	100·0

Primary energy consumption in the Soviet Union and East European countries: 1950–64

	Solid fuels	Natural gas	Oil	Nuclear and hydro	Total
		(In million tons coal equivalent)			
1950	350·3	11·8	60·0	5·9	428·0
1955	539·4	18·8	99·7	10·9	668·9
1960	681·6	79·2	177·4	22·8	961·1
1964	672·7	162·6	256·1	10·4	1,101·9
		(In percentages)			
1950	81·8	2·8	14·0	1·4	100·0
1955	80·7	2·8	14·9	1·6	100·0
1960	70·9	8·2	18·5	2·4	100·0
1964	61·1	14·8	23·2	0·9	100·0

Since 1964, moreover, demand for coal in Europe, in particular, has steadily declined at a broadly average rate of some 20 million tons a year, and this rate shows no sign of slackening, at least in the short-term future. It is noteworthy, nonetheless, that in the United States, where the coal industry has been extremely successful in reducing its labour force and adopting highly mechanized and scientific mining techniques, its market outlets have increased substantially and it has good prospects for meeting the challenge from nuclear power, at least in the short-term future. Thus, a recent American estimate put the amount of coal likely to be consumed in American power stations by 1987 at over 550 million tons out of a total of 1,270 million tons of coal equivalent; this compares with a coal consumption figure in this sector in 1966 of only 253 million tons. But even in the United States coal's ability, notwithstanding its extremely low costs of production, to withstand the formidable challenge from nuclear power in the medium and longer terms is open to serious doubts. One eminent authority, Mr R. H. Gerdes, President of Edison Electric Institute, recently gave the following possible pattern of new electrical power generation in the United States during the rest of this century:

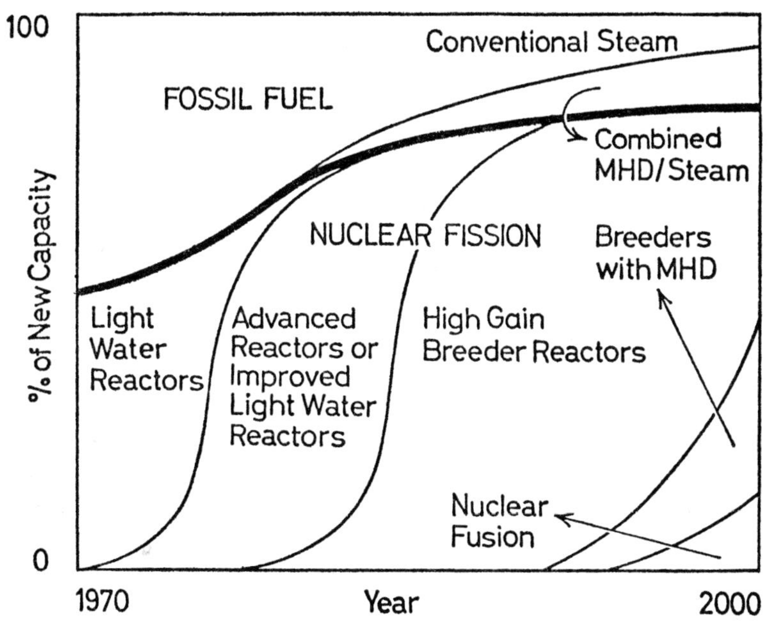

Figure 4. Possible pattern of new electrical power generation during the rest of the century in the US (after R. H. Gerdes, president of Edison Electric Institute)
Source: New Scientist, 19 August 1968, p. 431

The general decline of coal, taking the world as a whole, has been due in the main to its relatively higher costs of production. This has been particularly the case in Europe, although social factors have also played an important part. In the United States and the Soviet Union, where coal can be produced cheaply and competitively, and where output, in absolute terms, is still rising, other fuels are, nevertheless, dramatically increasing their share of the overall energy market. The explanation for coal's relative decline in these countries is to be found in two main reasons: its poor performance in terms of specific consumption and the much greater amount of labour required to produce it.

The most salient point to emerge from Table 7 is, clearly, the tremendous upsurge of the oil industry, particularly in Europe and Japan. The key to this development can be found to a large extent in the extraordinary growth in the fuel requirements of the transport sector, which has been a feature of all industrialized societies since the end of the Second World War. Pending the development of a competitive and commercially viable electric battery for cars and vehicles, this is a market which oil has made almost entirely its own, in the air, on land and at sea—the only obvious exception being the electric train. While this has provided the oil industry with a large and virtually completely captive market, it has also had the effect of multiplying the availability of heavier oils. From a small excess in the years immediately after the war, these heavier or black oils have come to represent an increasingly large element in oil companies' turnover and revenue. The result has been the tremendous expansion in oil usage by industry for heat generation. At the same time oil and, more recently, natural gas, have become the main sources of feedstock for the chemical industry which, in the course of the last decade, has become, as we have already had occasion to see, one of the oil industry's most important and intimately related consumer sectors. More recently, indeed only in the last two years or so, the oil industry has looked on while a new and even more formidable competitor, in the shape of nuclear energy, has entered the power generation lists. Although nuclear power plants are, at the present time, still few in number, largely unproved technically, economically and commercially, and still give rise to numerous question marks as to their future role, safety and viability, there can be no reasonable measure of doubt about the fact that, in the electricity generating sector at least, they will in the longer term represent a powerful challenge to oil, while at the same time threatening at least one and possibly both the last two important bastions of the coal industry—the coke ovens which provide the coke for the steel industries' blast furnaces and coal-fired power stations. The battle between oil and nuclear power, or perhaps it would be more accurate to speak of oil and electricity, seems likely to be highlighted by energy developments during the next two decades; at present the trumpets' sound is muted but there is a promise of bitter conflict in many markets where oil at present holds a large or dominant share of the power generating market. In the case of coal, the economic struggle is virtually already a thing of the past, since few, if any, leaders of the world's electricity industries hold out the hope that coal will be able to match nuclear power in cheapness, convenience or reliability. But the big problem that faces the governments of the larger coal producing countries is no longer one of determining cost characteristics or drawing up comparative scales of costs or values, but rather one of assessing the right time and moment to phase out the coal industries and to ring in the energy industries of the future for reasons not only of security of supply, regional wellbeing, regional manpower redeployment and redevelopment, but also, and here the problem arouses and obtains, it must be admitted, a suddenly more immediate and meaningful response of immense and undeniable political significance in terms of electoral and party votes. For if, as seems probable on the basis of experience to date, the social problems of the regression of the European coal mining industry are to be satisfactorily and humanely resolved, it will be due at least as much to the fear of electoral chastisement as to real concern at the plight of those whom the forward surge of the new technological age has rendered economically obsolescent.

SECTION III

THE COSTS AND BENEFITS OF CONVERSION

The economic impact of the new energy industries

In a global sense, the expression 'new energy industries' means, in fact, nuclear power. The oil industry, while it has undoubtedly multiplied in a remarkable and highly impressive manner in the last two decades, has been with us for more than fifty years both as a source of energy and, more recently, as a feedstock for the chemical industry with which, as we have already has occasion to see, it has become intimately associated. Already before the Second World War, in the United States, oil was making inroads into coal's consumer markets. The same process in Europe and Japan was delayed by the high transportation costs, the relative scarcity of oil and its unattractive price to industrial consumers. Oil's potential, on the other hand, was fully appreciated and its importance for the transport sector obvious for all to see. So much so that the Germans, faced with the acute problems arising from a shortage of oil in the Second World War, spent vast sums on research and development on extracting oil from coal despite the fact that the production of one ton of oil in this way required the sacrifice of 5 tons of coal. By 1943 German hydrogenation plants were producing 3 million tons of oil for German military needs.

More recently, natural gas has hit European headlines as first Italy and then France, Holland and Britain, in the North Sea, have discovered sizeable natural gas deposits. Here once again, it is often claimed that, as in so many other fields, Europe has followed where the United States has long since led the way; in this connection it must, however, be remembered that the natural gas industry in the US arose directly out of its large domestic oil industry. Already, by the early 1960s, natural gas had come to account for approximately one-third of total United States energy consumption—a proportion which it is never likely to approach in Europe. Nevertheless, despite its relatively small contribution to Europe's overall balance, the prospect of substantial natural gas supplies is one upon which many industrialists have seized avidly. That industry is fully alive to the attractiveness of natural gas was proved recently by the publication in 1968 of a report by the Chemical Industry's Association of Great Britain, which concluded, characteristically enough, that greater benefits would accrue to the British economy from supplies of cheap natural gas to the chemical industry than to any other industry. This report stated that the chemical industry at present uses petroleum feedstock for the production of synthesis gas for manufacture of ammonia and alcohols; olefines, such as ethylene and propylene; and acetylene. It was estimated that the industry might in 1970, for the manufacture of ammonia and alcohol, use a total of 320 million cubic feet per day of natural gas. This quantity of gas would represent a saving of 2·6 million tons per year of naphtha. Capital expenditure necessary to convert plants from naphtha to natural gas would be of the order of £2 to £3 million. It was believed that the first converted unit might be ready on stream twelve months after matters such as quality of gas, delivery pressure and price had been agreed, and all plants converted within a further twelve months.

The report went on to say that there were three main uses of natural gas as a fuel. First, in processes where the quality of the fuel is important as it affects the quality of the resulting product, for example, in lime burning—this was known as 'process fuel' (Chemical Industry Association (C.I.A.) members could use 50 million cubic feet per day in such processes replacing other fuels); secondly, in gas turbines to supply power for compressors, together with by-product steam; and lastly in the firing of furnaces and in operations where close control of heat input is necessary. The report stated that 420 million cubic feet per day of natural gas could be used as a fuel, in replacement of oil, coke and coal. Conversion of equipment, it was claimed, could be completed at a rate that would enable the industry to take its first delivery of gas within six months of the settlement of the price.

Finally, the report summarized the advantages which the chemical industry would derive from the general use of natural gas as follows: the use of natural gas to replace naphtha as a chemical feedstock would result in a reduction of import of naphtha. 320 million feet per day of natural gas is equivalent to 2·6 million tons per annum of naphtha, and this substitution would save about £20 million annually in imports; the use of natural gas would put the British ammonia and fertilizer industry in a more competitive position. This would facilitate a reduction in imports and an increase in exports of fertilizers and ammonia products; insofar as natural gas can be made available to replace fuel oil in power stations, its use by chemical manufacturers for the combined production of steam and power is a better use of national resources than for production of electricity alone. The overall thermal efficiency achieved because of the balanced use of steam and power averages 65 to 75 per cent, compared with an expectation of about 40 per cent in new conventional power stations and substantially less in existing stations; in contrast to domestic consumers, a steady load could be built up rapidly at much lower capital cost. This would enable the North Sea Gas reserves to be exploited to the best advantage of the nation as a whole.

Examples abound of countries in which the discovery of new cheap sources of energy have transformed national economies. Thus, the oil-fields of the Middle East are rapidly revolutionizing the industrial and social structures of this part of the world. Elsewhere the changing locational pattern of refineries and the industrial growth and expansion to which they give rise has stimulated the economic development of completely new regions and towns, ranging from Tokyo Bay to Sicily, the Mediterranean coast of France and Rotterdam. In the United States vast development schemes like that of the Tennessee Valley Authority owe their success very largely to the provision of large quantities of cheap energy. One of the most interesting schemes, however, is that of the projected expansion of hydro-power in the Siberian provinces of the Soviet Union. This includes the construction of such giant plants as the Bratskaya hydro-electric station on the Angara river, scheduled to reach a total installed capacity of 4,500 MW, and the Krasnoyarskaya power station on the Yenisei river, planned to reach 6,000 MW.

Nuclear power for a new age of plenty?

It was H. G. Wells who was the first, in 1914, to predict that the advent of nuclear energy would set the scene for a new age of plenty, leading to a profound transformation of man's social and economic organization and environment:

> From the beginning to the ripening of that phase of human life, from the first clumsy eolith of rudely chipped flint to the first implements of polished stone, was two or three thousand centuries, ten or fifteen thousand generations. So slowly, by human standards, did humanity gather itself together out of the dim intimations of the beast. And that first glimmering of speculation, that first story of achievement, that story-teller bright-eyed and flushed under his matted hair gesticulating to his gaping incredulous listener, gripping his wrist to keep him attentive, was the most marvellous beginning this world has ever seen. It doomed the mammoths, and it began the setting of the snare that shall catch the sun.

We have up to the present time been inclined to look upon nuclear power as a valuable contributory, and ultimately dominant, source of electric power. It is an understandable reaction, even though significant progress has been made in such varied fields as the use of isotopes in medicine and in industry, the use of new materials intimately connected with the development of atomic power stations, and the application of nuclear power for marine propulsion. Even within the field of electricity generation, nuclear power has been looked upon essentially as an alternative source of energy, albeit an attractive one, to other more traditional fuels. While this caution may be justified at the present time, there are great hopes of a major breakthrough in nuclear costs in the not too distant future—a breakthrough that would not simply consecrate nuclear power's advantage over other fuels in this field, but would open up new vistas of virtually unlimited power at prices that are a fraction of those we pay today.

It is a well-known fact that as a general rule costs per unit of electricity sent out decrease as the size of the power station increases. This has been the fundamental economic truth behind the great surge forward in the postwar era to ever larger generating sets, from the early postwar 150 to 250 MW sets, to 500 MW and, more recently, to 1,000 MW sets and above. These economies of scale are considerably more pronounced in the case of nuclear power stations than for conventional plants. In a fascinating article, entitled 'Low cost energy: a new dimension', published in 1968, Dr Phillip Hammond, who is the Director of the Nuclear Desalination Programme at the renowned Oak Ridge National Laboratory in the United States, recounted that, at the new 3,100 MW Tennessee Valley Authority nuclear power complex under construction at Brown's Ferry, the division of costs would be approximately 50 per cent for capital costs, 12 per cent for operation and maintenance costs, and 38 per cent for fuel costs. In the larger giant nuclear power station complexes of the high-gain breeder variety of the future which, he believed, would go up to 40,000 MW in size, these fuel costs would fall to virtually zero, thus reducing overall costs by a third. This process would, moreover, enable the whole design and structure of future power stations to be simplified, thus leading to still further savings.

With cheap energy plentifully available, the only problems the world would face would be those of locations for these monster power complexes and the use to which these inexhaustible supplies of cheap power would be put. It is at this stage that the prophecy of Wells of half a century ago becomes so relevant. Cheap nuclear power can be used to drastically alter nature herself, by blasting out new harbours in areas previously regarded as unapproachable or turning the course of mighty rivers, such as the Nile or the Niger and fertilizing vast new tracts in central Africa, or by reducing the arctic conditions of Siberia or the far north of Canada to more hospitable temperatures. These are projects of immense promise and potential value to the six thousand million souls that are expected to crowd this small earth by the end of the century.

This however, is not the place to peer into the future or conjecture what may one day be possible. We have striven deliberately to remain within the context of contemporary life and present-day economics. And yet, it is tempting to yield to at least one short quotation from those prophets of the untold future wealth and power that will spring from energy's metamorphosis from its position today of a valuable and indispensable source of heat and power to one in which it promises to become the ultimate raw material, a new philosopher's stone, that will roll away forever the penury of the past and roll in a new age of plenty:

I see the typical energy centre in a large grid as a floating complex some 15 km offshore. It would be located, where possible, in water at least 100 metres deep in order to reach the colder bottom waters below the mixed zone of the sea. This very cold water increases the power station's thermal efficiency. As the bottom waters are rich in marine nutrients, the 40 million cubic metres of warm water per day discharged from the condensers would support a large fish population.

We shall have occasion to look again at this link between fish farming and power station discharges later in this section. Passing on to a broader and more general theme, Dr Hammond concluded his article with this optimistic, wellnigh utopian passage:

The technical achievement of abundant cheap energy will be of primary importance to the extent that it can be applied to man's pressing needs—needs that are, in fact, basically social in origin. The task of predicting exactly what these needs will be or how our technology can be applied to them is far more difficult—and yet more important—than forecasting technology itself. The difficulty arises because the social problems to be solved do not yet exist—at least not in the context to which our technology can be related...The achievement of unlimited cheap energy and the elimination of 'have-not' regions should lead to a large number of such fixes [i.e. situations], most of which we cannot hope to predict. The new technology and the social changes thus generated will undoubtedly create some new problems for mankind in the course of solving others.

There is a conclusion which seems justified, however, in spite of our ignorance of the details; namely, that as energy consumption grows larger and yet less costly, mankind will exercise increasing degrees of control over his environment. And as environmental control is essentially synonymous with wealth, so kilowatts become, in effect, the medium of exchange. If man and his technology can somehow 'fix' enough of his social ills, he has the means to become wealthy indeed in the future.[1]

[1] See article in *Science Journal* of January 1969 by Dr Phillip Hammond, entitled 'Low cost energy: a new dimension', pp. 34–44.

The influence of nuclear development upon the industrial framework

After these exciting and glamorous glimpses into a future that is not, when all is said and done, alas, for tomorrow, it is time to come back to the sober reality of the present day. The principal immediate attraction of nuclear power—assuming, of course, that it lives up to its promise of generating cheap electricity at a cost of say, 0·40 pence per kWh or less—lies in the prospects which its development has opened up for producing energy at steadily falling prices as a result of continued physical, scientific and technological advances. Nuclear power, by breaking the previous umbilical connection between energy production and energy transportation costs, has come to represent a factor making for lower costs and consequently a greater measure of price stability in the overall contemporary energy equation. As such, it is in the process of becoming (and, indeed, in some cases has already become) a major element of policy for regional development by means of its ability to attract highly energy-intensive industries and, in the process, create new centres of industrial activity and employment. One problem which arises immediately in these circumstances is that of discrimination. Thus, in most cases, during the early period of operation of nuclear stations at least, it is most unlikely that the kilowatt of electricity produced from a nuclear origin will be differentiated in distribution networks from electricity derived from other sources. Where a policy of this kind is pursued, with nuclear plants largely or completely integrated in an existing electrical power network, the development of nuclear energy can only come to exert pressure on the price of electric current through the achievement of cost reductions as a result of the lower production costs of nuclear kWh. This inevitably looks like being a slow process and it is precisely in order to be able to take immediate advantage of the promise of cheap nuclear electricity that big energy consuming industries, such as aluminium works, are pressing hard for governmental authorization to construct their own power plants, independently of national or regional utility boards or organizations. While nationalized electricity industries have understandably tended to look upon such proposals with a jaundiced eye, such are the pressures arising on Governments from industrial and regional considerations, and through them on public utilities, that a number of joint ventures have already resulted. Among projects of this kind are those in the construction of a 700 MW plant at Tihange, in Belgium, where some 200 MW will be used to supply cheap electric power to an aluminium smelter and another at Ludwigshafen, in West Germany, which will channel part of its production of electricity to the chemical firm of Badische Analin. In the United Kingdom, there is the proposal for a joint Atomic Energy Authority—British Rio Tinto venture to provide power for a new isotope separation plant to complement the facilities at Capenhurst as well as to provide cheap power for an aluminium plant.

Quite apart from specific ventures of this kind, the growing interconnection, on the one hand, of grid systems between various producing areas or countries with different consumption or utilization routines, and the rapidly expanding possibilities and the increasingly ambitious schemes in transporting electric current over high-tension systems for distances of up to 500 km or more without incurring excessive additional costs per unit of electricity, on the other hand, have done much to eliminate the traditional cost discrepancies between high and low electrical energy consumption areas. At the same time the growing possibilities for substitution between different forms of energy are enabling industrial consumers to switch easily from one type of fuel to another according to price and suitability. In addition, the tremendous advances made in the last two decades in fuel valorization and the much lower specific requirements of energy that have resulted from it, have made the physical source of energy supply of far less importance in the choice of industrial location than ever before. At the same time other constituent factors in the production process, such as differences in transport costs for non-energy raw materials, transport costs for the finished products for the industry in question, the availability of adequate numbers of skilled or trained workers, and the availability of land, have lost none of their importance. To this must be added the need for water for cooking or cleansing purposes over a wide range of industries. From this it seems that we may conclude that, for the short and medium term at least, the effect of the advent of nuclear power and the cheaper electricity deriving from it, will have its main economic impact in the form of lower costs to a limited number of big consumers of energy, notably aluminium industries, with a general level of energy consumption of 15 kWh per kg, and the ferrotungsten, electrolytic, zinc and chlorine industries, with energy consumption rates of 8·4, 3·7 and 3·3 kWh/kg respectively. Elsewhere its role in determining or influencing the choice of location in industrial implantations seems likely to be limited in the light of the diminishing differences in energy prices between regions and, conversely, the enhanced significance of other locational factors. Above all, it must be recognized, albeit perhaps with a tinge of regret, that the economic advantages that arise from major industrial conurbations, e.g. the Ruhr, the Midlands, are such that despite their frequent contradiction and conflict with desirable or sometimes simply modest, human and social aspirations, they are able to exert compelling pressures upon workers to move to already over-crowded areas rather than introduce new industrial enterprises—almost inevitably with State or local subsidies—to establish themselves in regions that have become industrially or economically desolate with reserve pools of ageing, disillusioned, but still potentially active and useful men who have become victims of the modern technological age.

Born into the early post-war era, civilian nuclear power has enjoyed a relatively rapid growth and is now approaching the end of its industrial adolescence. But much in the same way as we cannot yet see with any real degree of accuracy what effect the development of space research or new generations of computers will have on the behaviour patterns of an increasingly industrial and technological society, so the future impact and significance of nuclear power remains uncertain. In this connection, it seems logical to consider, in the first instance, the effect of nuclear power on the construction of industrial plants and equipment used in the energy and, more especially, the electricity producing industries. The tendency towards the construction of bigger power stations, with a view to reducing the unit cost of electricity produced, is a well-known and near universal phenomenon which has been with us for several decades and which applies to all types of conventional, as well as more modern and sophisticated, power plants. The undoubted economic advantages that result from building larger power stations are even more substantial in the case of nuclear plants, by virtue of the significantly greater reduction per unit cost in relation to size. One of the most important technical size advantages for nuclear plants stems from the share of fixed expenditure in that part of total expenses arising from the surface area rather than sheer volume in the construction of a nuclear power station (i.e. the biological shield). In conventional power stations, on the other hand, development of larger boilers raises, once certain dimensions are exceeded, problems of heat distribution of such complexity that the cost of the additional equipment required for their solution more than offsets the value of any cost reductions which their solution may actually bring about.

It is a fact nevertheless that the large amounts of money now being spent in research on the feasibility and optimization of larger nuclear production units would also benefit conventional plants. Over and above units of 600 MW, the successful operation of steam generators, the distribution of electricity and the avoidance of energy losses are still presenting considerable technical problems. Since 600 MW is well below what is now generally regarded in electricity industry circles as the economic optimum, there can be little doubt that this size obstacle will be successfully overcome. The first signs of this are, indeed, already apparent in the United States and, to a somewhat lesser degree, in the United Kingdom. It would, however, be wrong and misleading to suggest that this could lead to virtually unlimited increases in size of classical conventional units.

Less encouraging from the economic point of view is the fact that, for commercial and industrial reasons, the present state of knowledge of nuclear power plant construction does not encourage the construction of serial plant building, which would have the effect of offsetting to some extent the advantages accruing from the trend towards large plants. The many different types of nuclear plants and reactor types, the big margin of potential improvements that remain, and the numerous different ways in which this might be achieved provide unanswerable economic, technical and physical arguments in favour of diversifying types and models of nuclear power stations to be built, operated and developed by the responsible electricity authorities. At the same time, the inherent advantages for the electricity producers in being able to operate a reasonably homogeneous range of plants, allied to the pressures exerted by governments or local authorities to prune back on research and development, as well as on operational expenditure, through lower investment costs as a result of serial construction—albeit at a price of higher costs in the longer term—will in all probability, particularly in the present difficult economic circumstances in many leading industrial countries, encourage a number of electricity generating bodies or companies to try and reach some form of compromise between technological innovation and experimentation, on the one hand, and the advantages of operating proved reliable and reasonably standardized equipment and units on the other.

In the field of transport of electric current, the increase in the size of new power stations, whether as a result of a general trend towards bigger plants or as a consequence of technological improvements in nuclear research and development, is bound to have a profound effect upon the capacity of electricity power networks. The first feature is likely to be an acceleration in the construction of larger high tension transportation lines of 380 kV and above. Thus, some American utility companies are already believed to be carrying out experimental trials on the feasibility of 765 kV lines. Another significant feature which is emerging, particularly where there are geographical groupings of a number of smaller countries, is the growth of inter-connecting national grid lines. This is made necessary by the fact that for many of these smaller countries the maintenance or observation of the security rule that no single electricity producing unit should exceed 5 per cent, or at the absolute outside 10 per cent, of the total installed network capacity is becoming increasingly difficult as well as uneconomic to observe. The inter-connecting of several national grid lines provides a convenient solution to this problem as well as assuring a wider and better regulated market for the baseload output of electricity which the new power stations will be called upon to provide. Thus, in the electricity generating sector at least, planning of production from nuclear power plants on a multinational basis, in the best economic interests of all countries concerned, can be expected to grow.

The introduction of new materials

The development of nuclear power for both military and civil application has given rise to a demand for new steels, alloys, and other materials with particular or special qualities. These include materials that have a high neutron-absorption capacity, such as cadmium, hafnium or europium; or the reverse, i.e. a high neutron rejection capacity, such as zirconium; or highly effective corrosion resistant materials, such as graphite for nuclear power stations. The use of these new materials has not, of course, been limited to the nuclear engineering side and their utilization in many other technologically advanced industries has grown extremely rapidly. In this way, ceramic materials with high densities and strong resistance, whose production today on a commercial scale can be ascribed almost entirely to the research work carried out on their potential use and application in the nuclear field, are now in wide use in mechanical construction, electronic and metallurgical industries. Similarly, the research work that is currently in progress on beryllium is confidently expected to enhance greatly the prospect for its use in other fields, particularly for jet engines (due to its light weight and remarkable resistance to high fusion temperatures). Examples such as these serve to illustrate the impact which the growth of the nuclear industry is having over a whole range of industries by means of the developments of new materials and processes that are also of benefit to conventional industries. This remains a comparatively new field, and the vistas that have been opened up particularly by the work on fast breeder reactors and experimental work on thermonuclear research are enormous; together, they are pointing the way to the commercial use and valorization of a range of materials which have until recently been, in a good many cases, little more than laboratory curiosities.

Everything of course in the nuclear field is not new. It needs, perhaps, to be recalled that the construction of nuclear power stations requires the participation and contribution of more than 100 enterprises of the conventional industry sector, either in the shape of direct suppliers or subcontractors. All of them, however, have had imposed upon them standards of quality, precision and reliability that are exceptionally rigid and strict and that require at the same time a co-ordination between all the companies or enterprises concerned that goes far beyond the ordinary norms of everyday industrial activity. The manufacture, control and processing of nuclear fuels have, for their part, given rise to yet another series of highly specialized plants, responsible for the operating of new techniques and the development of a new type of know-how.

Once again—as in the case of new alloys and materials—these new techniques of manufacture and preparation of nuclear fuels have found a ready application in the non-nuclear field. Thus, the work done in nuclear laboratories on increasing resistance to corrosion, which is required from construction materials used in nuclear power stations and on fuel-cladding materials and moderators to control the movement of neutrons in graphite, have enabled enormous progress to be made in achieving a better understanding of the process of corrosion. This, in turn, has led to the development of new and improved types of graphite, stainless steel and zirconium alloys.

All of these developments have compelled those firms or enterprises that wished to keep abreast of the times, and to play a part in the new technological age, to adapt themselves to the new situation brought about by the industrial, economic and scientific spin-off from nuclear power, by creating adequate research departments, specialized competent technical divisions, new management skills, and an entirely new structure of framework to promote co-operation and participation between enterprises. All this, although still perhaps a little unspecific and unclear—indeed the greater the flexibility the better at this stage—constitutes a highly positive, if not invaluable contribution by the industries associated, in a broad sense, in the nuclear field, to the modernization and continued advancement of all industrialized societies. Nowhere is this more truly apposite than in Europe, where the continual harping upon an industrial or technological inferiority complex *vis-à-vis* the United States, often magnified out of all proportion to the realities of the situation, has recently shown signs of erupting into a veritable orgy of self-denigration. Thus, there have already taken place, in the field of nuclear industries in Western Europe, albeit largely as a result of pressures of competitive necessity rather than of firms' own volition, several regroupings between enterprises of somewhat less than international dimension. Such mergers, while they remain of a relatively modest size and stature, have made a considerable step towards the ultimate objective of creating European industrial concentrations of sufficient dimension to enable them both to meet the challenge of overseas competitors in European markets and to be able to compete with them on equal terms in world markets. But, if this is the objective, it must nonetheless be recognized that there is still today a serious risk of the technical colonization of Europe by the leading American nuclear enterprises either through the channel of licence agreements, or the whole or partial take-over of European firms. The only way for Europe to avoid such a subordinate situation in the nuclear as well as other advanced technological fields, is through the creation of big autonomous enterprises that have the potential and the resources to remain economically viable. Such a policy, moreover, while desirable in that it would ensure that Europe would maintain its own voice and identity in a leading industrial sector, would have the additional advantage of encouraging European physicists and engineers to remain and work in Europe and so contribute to the plugging of the brain drain of European scientists and technologists to the United States.

It has been argued with, it would appear, a good deal of conviction and success that if the countries of Europe were to abandon their attempts to make the necessary financial and monetary sacrifices which are necessary if they are to develop in any real meaningful way the key industries that have come to characterize the modern technological era, such as aeronautics, space and satellite telecommunications, computers, automation, and nuclear power and thermo-nuclear physics, they would before very long find themselves reduced to secondary zones of reserve scientific manpower. In such a situation those most highly qualified would emigrate and the commercial and economic value of providing higher qualification and training for the remainder for fields in which Europe would no longer have a satisfactory or worthwhile stake, would soon be called into question. Europe would become a training ground for competent technicians capable of operating, at ground level, an industrial apparatus in process of continual change and development as a result of external, i.e. American, expertise. European enterprises would concentrate their efforts and their attention on goods and equipment less dependent upon continuing technological innovations and surrender the production of major new forms of equipment and new consumer goods—since the latter is inextricably linked to the former—to those countries able and prepared to

take the necessary steps to adopt and implement the required investment policies and organizational measures.[1]

The success or failure of these efforts to promote the growth of co-operation and industrial concentrations in the nuclear field may well have an effect which, if not decisive, will, nevertheless, prove to be of immense significance for the future industrial and technological potential of Europe in its confrontation with competition from the United States, the Soviet Union, Japan, and, in the longer term perhaps, from China. It is a fortunate fact that it is in the nuclear field that Europe's leeway *vis-à-vis* the United States is the smallest; not only have European countries, notably Britain and France, developed their own nuclear systems, i.e. the gas cooled reactor systems, but, above all, they can look forward with a considerable degree of confidence to the outcome of their intensive research efforts in the field of fast breeders. At the same time the omens for co-operation between these three countries most closely concerned cannot be regarded as being particularly favourable. At the moment each one of these three countries is pursuing, independently and with the apparent conviction that its own national brand of fast breeder offers the best prospects, its own objective. The result is a tripling of the costs and a triplication of the research work and effort required. A European failure in this field, where the prospects are still reasonable, due to national rivalries and jealousies, would be a disastrous body-blow to the development of an independent European industrial and technical infrastructure.

But whatever the gravity of its industrial or technological aspects, it is the human and social considerations which are most obviously in evidence. In countries with little or no indigenous energy production no such problems arise. But elsewhere, particularly in countries with large but, at the same time, relatively high-cost coal industries, these problems are, as we have already seen, still acute. If the time of reckoning in the United States is still some ten years or so ahead, as a result of the extremely low cost at which American coal can be produced, and the problem has already been largely resolved in countries like Belgium or Holland, it remains a very real, current and pressing problem in Britain, France, Japan and Germany. Nevertheless, the contribution that nuclear power will be called upon to make—and by this we must understand both its rate of development and the parts of the country where nuclear plants are to be located—must be decided as soon as possible in the best human social interests of all three countries concerned.

[1] In this connection, see notably J. J. Servan-Schreiber's *Le défi Américain*, Denoël, Paris, 1967.

What conclusions can we then draw about the economic implications of the rise of the new energy industries? There are, I think, five main conclusions. First, the rise of oil and natural gas, but even more especially the emergence and vast potential promise of nuclear power, have, among them, abolished the spectre of an energy shortage for an indefinite future; second, they hold out an assured promise not only of secure but also of relatively cheap energy supplies in all industrialized areas either directly or through their reducing effect on the prices of other fuels; third, this development holds out the possibility of a fundamental reorganization of industrial and regional patterns, whether this is desirable for social, economic or strategic reasons; fourth, while each new energy industry brings with it a valuable off-spin in new industries able to use it as fuel or feedstock, i.e. petro-chemical industries, the use of naphtha by the gas industry for gas making, the prospect of extracting food from oil, nowhere are these prospects so rich and almost beyond our present assessment capacity as in the nuclear field; and fifth, an obvious point, but one which has been insufficiently stressed, the fact that the effect of these changes upon our corporate society, upon both its industrial, social and, indeed, personal aspects is bound to be very great indeed.

A case study in employment prospects

The problems involved in the previous paragraphs are of course of a universal application. But on a more limited, if nonetheless important, scale it is interesting and worth while to examine the detailed figures and conclusions of a study prepared by the Commission of the European Communities comparing the respective manpower requirements of nuclear and conventional power station construction programmes of comparable size.

On the assumption that the amount of generating capacity commissioned in the three years 1970, 1975 and 1980 would be of the order of 2,000, 5,000 and 9,000 MW respectively, it was estimated that total industrial turnover arising from the construction of nuclear plants, or, by way of comparison with conventional plants of similar capacity, in the immediately preceding years, expressed in $ million, would be as shown in Table 8.

If the turnover of enterprises is a yardstick of their activity and prosperity, there is also a generally admitted statistical link between global turnover and the volume of manpower employed. This relationship varies, of course, considerably, according to the degree of mechanization or automation of the industry concerned. The industry can be expected to move forward, in most cases regardless of manpower redundancy considerations as a direct function of the development of new techniques and

Table 8.

Comparison of turnover by equivalent construction programmes of nuclear and conventional power stations

(In $ million)

Industries	Nuclear power station programme 1969	1974	1979	Conventional power station programme 1969	1974	1979
Civil Engineering	38	90	160	62·5	156·25	281·25
Metal construction	201	471	845	87·5	218·75	393·75
Electrical construction	141	339	596	100·0	250	450
Non-ferrous metals	63·4	238	536			
Precision engineering	48·8	171	432			
	492·2	1,309	2,569	250·0	625·0	1,125·0
Conventional fuels				50·4	423	1,280
				300·4	1,048·0	2,405·0

the organization of new production processes. This tendency towards higher productivity will find a natural reflection in an increase in turnover per employee. Conversely, an industry with a stagnating turnover would tend to reduce its labour force.

In our present state of statistical knowledge and expertise, the data available, even on a limited, say European scale, is too insufficient and too imprecise to permit the elaboration of detailed or sophisticated ratios of turnover to manpower for industries. The Community experts took for the purpose of their study figures that were available for West Germany and, while admitting that the degree of industrial concentration and consequently the rational deployment of labour was a good deal more advanced there than in almost any other industrially mature country, nevertheless worked on the assumption that these figures could be regarded as being reasonably representative of the Community as a whole. If we take the figure for 1965, we see that turnover in the civil engineering sector for the month of June of that year was given as DM. 1,146 million; and the total number of people employed as 528,000. Annual turnover per number of persons employed in 1965 was then, somewhat crudely, calculated as being equivalent to

$$\frac{1{,}146 \text{ million} \times 12}{528} = \text{DM. } 26{,}000.$$

The comparable figures for the other industries listed in Table 8 were calculated to be as follows:

Metal construction industry	DM. 35,300
Electrical construction industry	DM. 32,400

Non-ferrous metals	DM. 70,800
Precision engineering industry	DM. 24,600
Conventional energy industries:	
Oil	DM. 372,500
Coal	DM. 20,600
Processing industries	DM. 45,600

The steady rate of increase in industrial productivity, which has been such a marked and significant feature of the post-war era and which seems likely to continue during an indefinite period of time ahead, will, inevitably and automatically, have as one of its consequences increases in turnover over a wide range of industry. It is, of course, equally obvious that this increase is unlikely to be uniform in all sectors of industry in all countries.

In attempting to assess the effect of increased productivity upon turnover, and hence upon employment, the Community experts assumed a rate of increase in productivity for the Community as a whole of 3·5 per cent. The combined effect, for example, of the trend towards building larger power stations and improved productivity in the construction industry will be to reduce the level of investment in real terms per unit of installed capacity; this must obviously play a part in assessing the total level of investment required for a new power station. In their study the Community experts consequently worked on the basis of new standard-sized stations of 500 MW sets, since the figure seemed an acceptable and reasonably accurate compromise between the traditional small production units and the modern trend towards plants of 600 MW or more. Thus, in the United States and in the United Kingdom, the responsible electricity authorities are already going well in excess of 600 MW, although in many cases this has consisted rather of coupling or tripling a number of 600 MW units rather than embarking upon the construction of one mammoth 1200 MW unit with all the enormous technical, economic and, above all, financial risks this would involve. For the purposes of calculating the compound increase in productivity by 1974 and 1979, the Community experts assumed a continuous increase of 3·5 per cent a year over a period of 12 years; for 1969 a similar rate of increase was assumed over a period of four years only. This gave an estimated rate of increase of productivity of 15 per cent for power stations built in the period up to 1969 and an average of 50 per cent for those built between 1970 and 1979. It followed that the turnover figures which were taken as a basis of calculation had to be increased accordingly.

The value added by each industrial sector in the manufacturing process is the cost or value of the processing applied to the goods which that particular sector has bought from its suppliers. The building of power stations and their fuel supply requirements create a large additional turnover by virtue of the enormous amount of accessory equipment they require, from raw materials for the cement industry to provide the concrete shielding to special steels and alloys. The Community experts estimated the total number of jobs that would have to be created in all related industries, for the objective of a target programme of 60,000 MW of nuclear capacity in the Community by 1980 to be realized, at 45,000 persons in 1969, 91,000 in 1974 and 176,000 by 1979. Comparable figures in the event of similar additional coal or oil-fired capacity being decided upon were also given:

	Coal	Oil
1969	32,000	24,000
1974	90,000	45,000
1979	220,000	82,000

Broken down into more detail, this meant that for the construction of 60,000 MW of nuclear capacity there would be 20,000 men employed in construction work, another 6,000 in ensuring an adequate fuel supply, and 18,000 in related industries fully occupied on this task: a total of industrial manpower therefore of 44,000 persons. The comparable figures for coal and oil-fired stations for 1969 and for all three forms of energy for 1974 and 1979 are set out below:

	Construction jobs (*a*)	Fuel Supply jobs (*b*)	Jobs arising from (*a*) and (*b*)	Total industrial manpower
1969				
Nuclear	20,000	6,000	18,000	44,000
Coal	14,000	6,000	10,000	30,000
Oil	14,000	200	10,000	24,000
1974				
Nuclear	36,000	15,000	38,000	89,000
Coal	25,000	40,000	24,000	89,000
Oil	25,000	1,200	18,000	43,000
1979				
Nuclear	65,000	37,000	73,000	145,000
Coal	45,000	122,000	52,000	220,000
Oil	45,000	4,000	33,000	82,000

From these tables, the Community experts drew a number of tentative conclusions. First, that for a projected nuclear construction programme of, say, 60,000 MW (as was the case for the Community) the level of investment would be substantially higher than in the case of conventional plants of a similar total capacity. It followed that the number of jobs created in industries investing capital, time and labour in the actual physical construction of power stations is bigger when nuclear plants are selected in preference to conventional power stations—the difference amounting to about 11,000 jobs by 1974 on the basis of an annual rate of investment and construction of 5,000 MW and 20,000 jobs by 1979 for an annual rate of investment and construction of 9,000 MW.

Secondly, the investment required in order to supply the fuel charge for the first annual tranches of 5,000 and 9,000 MW respectively, together with that required to provide supplies to the plants already in operation (i.e. 16,800 MW in 1974 and 50,800 by 1979), added up to a very much smaller number of jobs than a comparable coal-fired station programme. This was due to the fact that the relative price of coal was many times higher than that of uranium, and turnover per employee in the coal industry very much lower than in the nuclear industry: in terms of numbers of men employed—albeit much less gainfully—a coal solution would provide 25,000 more jobs in 1974 and 80,000 more in 1979. An oil-fired programme would give rise to the smallest number of jobs of all since oil is an imported product which requires little handling in what are normally largely automated refinery plants.

Thirdly, as far as indirect employment arising from the construction of nuclear power stations and the provision of their fuel supplies is concerned, the higher absolute value of nuclear investment, combined with the relatively high proportion of work done by conventional industrial firms on a sub-contracting basis, gave some advantage to the nuclear solution in this respect, i.e. by providing some 14,000 additional jobs in 1974 and about 20,000 by 1979. This advantage would, of course tail off quickly in the event of a complete stoppage or even a modest decline in the annual level of nuclear investment, due to the fact that turnover for related industries in a nuclear construction programme arises essentially from the original nuclear plant investment rather than from continuing fuel supplies. For coal, on the other hand, with its big labour force and consequently comparatively light calls on outside supply industries, the situation is reversed with the bulk of expenditure arising from the continuing provision of fuel supplies.

The Community study, while pointing to some valuable social considerations, made no attempt to delve into the economic problems that bristle formidably in any energy policy in which nuclear power plays a prominent part. This is not due so much to the uncertainty that admittedly still exists at the present time about the true costs of nuclear power and its ability to compete with other fuels, particularly oil, on equal terms, but above all to the social, regional and financial consequences of an accelerated and unnecessarily rapid rundown of traditional industries such as coal mining, with labour forces. It is to an examination of the cost of such a policy, and the economic gains and losses arising from it, that we shall turn in the second half of this section.

Energy as a source of food

A great deal of research work is being carried out by oil companies, notably British Petroleum, Esso and Shell, on extracting food from oil and natural gas. Not, before the gourmet becomes too alarmed, for direct human consumption, but to provide protein for animals. A recent article in the *New Scientist*[1] stated:

> It need cause no surprise that some sort of protein is available from methane. Indeed, it is sometimes said that there is hardly a material known of which some micro-organism cannot make good use. While some bacteria are totally dependent on a complex host environment, others have a built-in adaptability which enables a single species to use any one of over a hundred alternative organic compounds as its sole source of carbon. Many of them derive their energy by oxidizing, instead of carbon compounds, inorganic materials such as sulphur, hydrogen and combined iron.

The problem, therefore, was to find an organism capable of metabolizing methane, and, having done this, to develop a suitable extraction process and establish that the organism constitutes a safe and useful food. It has been estimated that the various processes that are at present being tested or are in use might be capable of producing one ton of bacterial protein from three tons of methane. As far as the food value of proteins obtained in this way is concerned, the *New Scientist* article commented:

> It has so far been established that they seem to be non-toxic (the experimenters, as well as their laboratory rats, have eaten them with impunity). The balance of amino acids and the vitamin content have not been published, but one of the researchers has hazarded the guess that an animal might be able to live on the bacteria exclusively. This guess is academic, of course, for the aim of all such work is to produce a source of high protein food which can be used to supplement a cheaper but deficient traditional diet.

The oil companies are not prepared at this stage to reveal detailed information about the economics of the process. The simple fact, however, that several of the major oil companies regard it as being of sufficient promise to maintain a considerable research effort in this direction is a good indi-

[1] See the *New Scientist* of 16 June 1966, p. 711.

cation of the potential value that they see emerging from this line of research. In April 1969 it was, moreover, announced that British Petroleum had in fact sold a licence for the production of a protein concentrate from oil to a Japanese company. B.P. is itself, moreover, constructing at the present time two commercial protein manufacturing plants, one at Lavera, near Marseilles, in southern France, and the other at Grangemouth in Scotland, with annual capacities of 16,000 and 4,000 tons respectively.

Another fascinating experiment in the same broad area of research is that of rearing marine fish in warm water effluent of electrical generating stations. In an article published in November 1968,[1] entitled 'Power stations as sea farms', Dr Nash, who is in charge of the experimental fish rearing plant at the Hunterston nuclear power station in Scotland, said notably:

> There are a number of ways in which marine fish farming might be carried on. One could repeat and improve on the basic productivity work of Gross and his associates who, more than 20 years ago, added fertilizers to certain Scottish sea lochs. An alternative possibility is to supplement the natural food of fish contained in the sealed arms of sea lochs, or in netted areas within the loch. Or the fish could be raised in tanks supplied with water from the sea. Or they might be installed in tanks at a power station, using the warm water discharged to lengthen the growing period, and fed entirely on waste, organic or artificial food, or on live specially cultivated foods.

Dr Nash went on to describe two trials held in 1966. The results in his own words were extremely encouraging:

> The first trial was held at Carmarthen Bay, a conventional coal-fired station in South Wales, and the second at Hunterston, a nuclear generating station in Scotland. These stations were chosen because they both employ a system of injecting a continuous low level of chlorine into the intake fumes to keep marine organisms from settling. The chemists at Carmarthen Bay had developed the continuous system in an attempt to find an economic and efficient level of chlorine addition. Their techniques had later been adopted at Hunterston. In the preliminary experiments no attempt was made to maintain a fixed temperature level as future development would have to accept the diurnal and seasonal variation of the discharge system. The results showed that both plaice and sole survived the low levels of free halogens to which they were exposed and they obviously benefited from daily feeding and attention, and the higher temperature range. Both the trials were begun in winter, and involved nine-month-old fish measuring between five and nine centimetres. A year later most of the fish of both species were of marketable size, that is 23–24 cm. The sole tolerated the conditions better than the plaice in that they withstood the higher summer tank temperature. This is to be expected, as their natural distribution is in warmer latitudes. Plaice, however, were distraught at temperatures above 19 °C and there were some mortalities at both sites. The extension of the growth period through the year enabled the fish to reach marketable size at least 12 months before the most advanced individuals in natural conditions.

[1] See *The New Scientist* of 14 November 1968, pp. 367–9.

Hunterston is a base load station, and with the agreement of the South of Scotland Electricity Board, it was decided to continue the feasibility study there. Four tanks were constructed, 14·4 m × 7·2 m × 1·0 m deep. These were ready to receive hatchery-reared sole in October, 1966, and a number of young fish, three to four centimetres in length, were safely transported from the Isle of Man hatchery. But the hatchery stock that year was suffering from the effects of unsuitable larval food, and there were high losses. At Hunterston, more than 30 per cent of the fish that died succumbed in the first week. Many more deaths must have gone unrecorded. By March only six per cent were still alive, but of these—not yet two years old—62 per cent were of marketable size.

Both the preliminary trials and the first field work proved the feasibility of using power station condenser discharges as an environment in which to grow certain species of marine fish, and indicate a feasible system on which a commercial enterprise may be founded.

Clearly, here again, the case is not yet fully proven and many problems still remain to be solved; it is nonetheless evident that there are enormous possibilities in this direction which will sooner or later, and probably sooner, have valuable commercial applications.

Some problems for the future

The advent of the new energy sources, and of nuclear power in particular, has thus raised new hopes and created new problems. The immense potential boon of unlimited cheap power is balanced by fears of its potential misuse and misapplication over the whole field of industrial and urban endeavour. It was to these problems that the Gif-sur-Yvette Conference of 1968, to which we referred at the beginning of this book, principally addressed itself.

The Gif-sur-Yvette Conference had five main points on its agenda:

(i) which forms of energy are likely to make the main contributions to the world's energy supplies during the next 50 years?

(ii) to what extent will the energy produced from these sources of supply permit the economic use of small and compact electricity generating units?

(iii) to what extent will it be possible to construct the big power station complexes of the future near to big consumer areas, or, alternatively, will it be necessary to build them near to the sea-shore?

(iv) what difficulties will the overland transportation of large quantities of electrical or primary energy give rise to?

(v) what will be the nature and the extent of the problems arising from ensuring adequate supplies of energy to the major urban centres of the future?

As far as the determination of the main form or forms of energy of the future is concerned, the Conference was almost unanimous in the view that, well before the end of the century, nuclear power would have established a dominating position in the world's energy market. This development was regarded as utterly inevitable; the only elements of uncertainty were the speed with which it would take place and the measures that could or should usefully be taken to control it. While we shall be coming back to this point it is instructive to summarize briefly, at this stage, the main conclusions of the Conference in this respect. Coal, it was felt, had little or no hope of being able to compete with oil or nuclear power. Oil, in its turn, would find the competition from nuclear power increasingly difficult to meet. The world-wide trend towards cleaner air and anti-pollution legislation was repeatedly emphasized as a major factor in favour of nuclear power. There was, it was admitted, the problem of disposing of nuclear waste but there was no lack of adequate or ingenious solutions, ranging from its immersion in steel containers to its expulsion by rocket to the moon or into outer space. As for balance of payments considerations, here again this was a point which, where it was appropriate, tended to the ultimate advantage of nuclear power.

On the second item of the agenda, namely the possibility of developing small electricity generating production units, on a commercially viable scale, the Conference held out few prospects of success. While technically the construction of small nuclear power stations was entirely feasible, economically it seemed likely to remain at a very great disadvantage when compared with the larger stations.

As far as the siting of the large nuclear power stations of the future was concerned, it emerged that there were a number of advantages favouring their construction by the sea. These advantages included some arising from the anticipated existence of a large number of big electricity consuming industries along the sea shores of the world drawn thence by their growing needs for imported raw materials. This applied with equal effect to conventional power stations which would depend on imported oil. Above all, however, and this was particularly applicable in the case of nuclear power stations, the need for large quantities of water for cooling purposes would be paramount. By way of illustration, it has been calculated that for a nuclear complex of 2,000 to 2,500 MW the amount of water required would be of the order of 80 m^3 per second for each tranche of 1,000 MW. The water requirements for a conventional power station of, say, $4 \times 1,000$ MW at 160 m^3 per second would be almost 50 per cent less. An industrial invasion of the sea-shore of this magnitude naturally raises many fears for the future. Not only is there a question of suitable sites but other no less important considerations such as the preservation of natural beauty spots, the interest of the tourist industry and the maintenance of reasonable stretches of unspoilt coast-line, are aesthetic and psychological imperatives whose importance must not be minimized. It is a situation which calls for the elaboration of a new economic geography within which regional and above all international co-operation and co-ordination will become matters of vital necessity.

The position may be eased to some extent by the continuing trend to build oil refineries at inland centres. The situation as far as refineries are concerned is, of course, somewhat different. Their need for water is both less voracious and less compelling and this provides them with a far greater degree of locational flexibility. The advantages of siting them nearer to the consumers of oil are clear, since it is much more economical to transport crude oil than it is to transport refined products; and while many new industries will gradually move out to coastal areas, the process will be gradual and many of the big traditional inland industrial conurbations will obviously remain. In these circumstances it is possible to envisage a situation in which inland power stations, usually of medium to medium-large size, will be mainly oil-fired, while the very big nuclear power stations will be located in peripheral areas on or near the sea.

On transportation, the point was made that one 730 kV line can transport as much energy as twenty 220 kV lines. With demand for electricity expected to continue to increase at a rate equivalent to doubling every ten years or so, the need to switch to high-tension systems was self-evident. There were, however, some technical and financial problems still to be resolved, such as the introduction of super-conductor cables, and the higher costs they involved. With regard to hydrocarbons, the use of ever larger pipelines was regarded as offering almost unlimited scope for improvement, and any further problems that might occur in this field were more likely to arise from discharging and stocking problems.

Finally, item five, the ability to ensure adequate fuel supplies to the urban centres of the future, clearly presents a colossal challenge in the field of organization. It is not only a question of one form of energy, since electricity, oil products and gas are all concerned. The solution suggested by the Conference was that of constructing special transport channels along lines not dissimilar from those of motorways or railways. The essential factor was to take the necessary measures, such as land reservation and compulsory purchase, in good time.

The problem is, however, easier to resolve in the case of inter-urban transport than in the cities themselves. Thus, in the older towns, the facilities for constructing 'energy' channels of this kind are extremely limited and immensely costly. In new towns and suburbs, on the other hand, the construction and planning of very large diameter pipelines for the common transport of water, gas, electricity and also, perhaps, telephone and telex cables should be strongly encouraged. Another possibility was the exclusive allocation to particular areas or suburbs of one type of fuel, say gas or electricity. This, it was rightly pointed out, would represent a major intrusion on the freedom of choice of the individual consumer; this was countered by the flat statement that it was part of the price that society would, increasingly and inevitably, have to pay for living in the super-industrialized and technologically sophisticated communities of the latter decades of the twentieth century.

The social and regional consequences of the rationalization of the coal industry

It has been argued on several occasions, especially by representatives of the Western European coalmining industries, that when account is taken of all the costs that arise in the process of reconversion of a particular area, the result of a massive campaign of pit closures will not necessarily prove to be a net gain to the national economy. Those who take this line of argument claim that account must be taken, in each case, not only of the costs arising from the closure of a particular mine, but also of those arising from the enforced stoppage of the operations of a number of related industries and other economic activities linked to the mine as well as the expenditure to introduce and establish new industries in the affected areas. But even this is not the end of the sum. Still to be added are the costs of the re-adaption and vocational training of reconverted workers and the premature retirement of a number of employees. Finally, it is argued, some thought must be given to the growing amount of short-time working or enforced idleness that is likely to occur among the men left in the coal mining industries as demand for coal continues to fall and the opportunities for their re-employment in outside industries become less and less attractive.

An evaluation of these costs is not a simple matter, and the conclusions in the most exhaustive studies made to date in this field of research are often of a conflicting, if not outright contradictory, nature. One of the first of these exercises was made in Germany in 1966 when the Ruhr Coal Producers (with, admittedly, a direct commercial interest in the outcome of the exercise) made a detailed survey of the costs that would arise from a reduction in production capacity—not of course necessarily the same thing as actual production—of 25 million tons a year for a period of three years. According to their figures, the annual costs arising from a pit closure programme of this magnitude were of the order of DM. 246 million. Where, however, the period of time during which many of the obligatory annual payments consequent to the closure of pits would have to be continued—ranging from two years to nigh on infinity—was taken into account, their sum total came, on the basis of currently payable rates of interest, to no less than DM. 2·1 thousand million. Moreover, account had to be taken not only of the costs falling on the coal industry itself but also of those which would have to be borne by the Government or local authorities. Some of these it was hardly possible to compute, but their sum total was expected to come to not less than DM. 439 million for each of the first three years, and DM. 2·7 thousand million in total.

When to these already enormous sums of money there were added still further costs (or losses) that could only be calculated in a very rough and ready manner, such as loss of business for the mining equipment manufacturers put at DM. 600 million a year, the loss of revenue for the railways and canals as a result of their loss of freight put at DM. 215 million, and the loss on taxation revenue payable by the pits in the process of closure put at DM. 100 million, the total sum at the end of the exercise approached DM. 6 thousand million or over £600 million.

Even this, however, was not the end of the story; it is not sufficient to close mines and provide early pension or retirement benefits, redundancy payment or retraining and reconversion credits. Work, in the form of new factories and, in many cases, a new local or regional infra-structure, has to be provided for redundant coal industry workers. Here again, it is extremely difficult to obtain very precise figures. A Belgian estimate, made in 1967, put the investment cost of creating one new job at between FB. 1 and 2 million or between £8 and £16 thousand. By way of illustration, it was stated that the closure of a mine employing 6,000 men would require a new investment outlay of between FB. 6 and 12 thousand million. A German estimate made at about the same time put the cost of creating one new job at DM. 100,000—broadly the same order of magnitude as in Belgium. A later, semi-official estimate, made towards the end of 1968, assessed the cost of creating 12,000 new jobs in the Ruhr area for redundant miners at DM. 4,000 million, equivalent to some DM. 300,000 per man.

The opponents of a massive reduction in coal output have consistently maintained that there are many other, more general economic considerations that must be borne in mind if the total cost or benefit to the community as a whole of a cutback in coal production is to be properly calculated. A first point with which they make considerable play is that the new sources of energy which are replacing coal are highly capital intensive and tend to place severe strains on the available amounts of capital. Moreover, and this is a second point of some significance, there is the fact that the construction of a new infra-structure, i.e. roads, air and sea communications etc., are more often than not a *sine qua non* of the successful reconversion of what were formerly essentially coal mining areas. This once again places the government or local authorities concerned under very great financial obligations. The result could well be an increase in pressures: thus, in fact, militating against the creation of new industries in development or backward areas. Finally, it is emphasized, a comprehensive and effective reconversion policy can only be carried out in an atmosphere of calm and on the basis of a carefully planned regional strategy. It should not and could not be achieved by means of decisions taken in a climate of precipitate political or economic expediency. New companies or enterprises created in an atmosphere of desperation would inevitably suffer grave structural weaknesses born of their improvised nature and the highly artificial conditions that give rise to them. So much so, indeed, that once the initial protection given them in the form of financial credits or tax reliefs was exhausted they would often be faced with problems as acute and as seemingly insoluble as those now facing the pits whose places they are designed to fill. In short, it is imperative to realize that a policy based on the large scale closure of pits under the pretext of providing new and cheaper sources of energy can result not only in dangerously increasing a given country's (or Europe's) dependence upon outside suppliers, but can also saddle it with extremely onerous and insufficiently appreciated financial costs.

As expected, a great degree of emphasis upon the financial consequences of the rundown of the coal industry, in terms of heightened regional and social costs, has also been evident in the United Kingdom, although the main source has proved, surprisingly enough, to be not the coal industry, but academic establishments. The explanation for this situation is perhaps to be found in the fact that the British coal industry is the only one in Western Europe whose leaders have not yet openly thrown up the sponge and accepted that the rundown of coal production and its replacement by other sources of energy is simply a question of time. To defend the continued existence of the coal industry on grounds of regional policy considerations or in terms of minimizing social hardship may be proof of an awareness of national needs or sound evidence of Christian virtues, but it is hardly an expression, when all is said and done, of confidence in the future of coal.

In an article published, appropriately enough, in *New Society* in September 1967[1] entitled 'Cheap energy; the costs of killing off coal', it was stated that the coal industry was confronted with two main issues: 'Is it true that nuclear power and natural gas will produce cheaper electricity than coal? And even if they do, will the national costs and benefits of their use justify their expansion in the consequent closure of the coal mining industry?'

Today, it would, most experts will agree, be confidently stated that nuclear power is generally, if not yet universally, cheaper, more economic and hygienically better and safer than coal. But the second part of the question still remains unanswered and unresolved.

In this article, the author, Mr Peter Odell, at that time Senior Lecturer in Economic Geography at the London School of Economics, dwelt with particular length on the fate of six collieries in the Durham coalfield whose future was intimately connected with the decision of the Minister of Power on the type of power station that was to be built at nearby West Hartlepool. The situation was that six modernized collieries in East Durham, each employing some 500 men, and believed to be capable of producing about 1 million tons of coal each at about 3·5 pence per therm, were exceptionally well placed to meet the requirements of the projected 2,000 MW power station. The cost of modernization of the six pits in question was not revealed but could, by analogy with modernized or reconstructed pits in other parts of the country, be safely assumed to be not less than £10 million. Because of the local market situation the coal produced at these pits could not be sold, profitably, or even economically, to consumers outside the East Durham area because of the high rail or alternative transport charges involved. In the event the Minister of Power gave his approval for a nuclear plant to be built. The consequence seems likely to be that the six pits will be closed, that 3,000 miners will lose their jobs, that yet another part of north-east England will be gradually depopulated and abandoned, and that the nearby 'new' town of Peterlee will be even more hardpressed to make life for its unfortunate inhabitants economically and socially worth while. 'If,' Mr Odell continued,

all the miners could immediately get alternative jobs locally, there would be no cost to the nation in the closures of the pits (this, of course, conflicts sharply with the views expressed in the Ruhr Coal Producers study). The economy would make a net gain from the nuclear power station of £2 million a year (the amount by which it would be cheaper to generate electricity from a nuclear rather than a coal fired power station) plus whatever other contribution the former miners made the economy in their new jobs. But this is impossible. A few miners may find employment locally; and rather more within a daily round journey of 50 or

[1] *New Society*, 21 September 1967.

60 miles (to Tees-side and Tyneside): but most will remain unemployed or be forced to move to more prosperous parts of the country.

This conclusion appears inescapable when one notes that the 3,000 miners in question will be the remnants of a present mining force in these pits of over twice the size. The men who are today being replaced by mechanization and automation will have taken the locally available jobs in an area which anyway has a consistently high rate of unemployment. We may, therefore, assume, that 80 per cent of the new displaced miners—some 2,400—will fail to find alternative work. Their redundancy pay, unemployment pay and other allowances will, by then, cost the country at least £1·8 million, i.e. almost as much as the direct annual saving on the nuclear power station.

Moreover, as the unemployed miners move away from the doomed area to find jobs there will be a lengthy period when the social cost will escalate rather than diminish. Existing local housing, education, social and health facilities will have to be written off and provided anew from the public purse (at higher capital cost) in the areas to which the ex-miners and their families move. The costs of this could only be quantified accurately with a good deal of research. But there seems no doubt that the annual costs, over a long period of time will exceed by several times the annual benefits to be derived by the nation from the nuclear power station. Hence Seaton Carew atomic power station is certainly not the best buy for the nation in the early 1970s. It will bring lower returns than from an alternative coal-fired station located to burn East Durham coal.

This uneconomic result of the decision to build a nuclear power station in Durham will be repeated several times across the country: in the North West at Heysham, in central Scotland, in South Wales and in the West Midlands. Each decision to go atomic and to close down marginally uncompetitive coal will quickly cost the nation millions of pounds.[1]

With the prospect of the adoption of a third nuclear programme in the United Kingdom of between 8,000 and 32,000 MW,[2] the overall cost to the country, on the basis of Mr Odell's calculations, could run to £15 to £20 million a year.

It is interesting, against the background of the Minister's decision in favour of a nuclear station, to look for a moment at the very strong views expressed by Sir Stanley Brown, the Chairman of the Central Electricity Generating Board, in favour of nuclear power and about the N.C.B.'s offer of coal to the new power stations at a delivery price of 3¼*d.* a therm:

It is also true...that even if the price of its coal were held stable at 3¼*d.* a therm for the next thirty years, a base-load coal-fired station at Hartlepool would cost over £1 m. a year more than the nuclear station which the C.E.G.B. wish to build here.

...The C.E.G.B. have told Lord Robens that they are ready to buy tens of millions of tons of coal at 3¼*d.* a therm but not for consumption at Hartlepool. The facts are that we are currently burning more than 60 m. tons of coal a year

[1] *Ibid.*

[2] Report from the Select Committee on Science and Technology, Session 1966–7, United Kingdom Nuclear Reactor Programme, p. 468.

at an average delivery cost of 4·7*d.* per therm as against the Coal Board's offer of 3 m. tons a year at 3¼*d.* per therm delivered to Hartlepool.

Of the 60 m. tons only 2 m. tons cost less than 3½*d.* a therm. About 12 m. tons cost between 3½*d.* and 4*d.*; about 13 m. tons between 4*d.* and 4½*d.*; more than 20 m. tons between 4½*d.* and 5½*d.*; and more than 10 m. tons cost over 5½*d.* a therm. Clearly, if coal can be made available at 3¼*d.* a therm it should be supplied instead of dearer coal and thus help to reduce the costs of electricity supply. It should not be used to prevent the construction of a nuclear station which would have the lowest costs of all.[1]

The Ministry of Power, for its part, clearly found the C.E.G.B.'s forward cost estimates more convincing than those put forward by the National Coal Board since the Minister, when announcing the Government's decision on the type of plant to be built, estimated the cost per unit of electricity from a nuclear plant at 0·52*d.* against 0·70*d.* for a coal-fired station, with a 75 per cent load factor in both cases.

In contrast, the essential core of Mr Odell's thesis was taken up and elaborated with some apparent conviction in a report prepared and published in the autumn of 1968 by the Economical Intelligence Unit entitled: 'Britain's Energy Supply'. While this report was commissioned by the N.C.B. as part of the campaign to try and force an independent enquiry into nuclear costs, its authors insisted that it had been drawn up on the basis of complete impartiality.

The purpose of the Economist Intelligence Unit's study was, it was stated, to assess the net economic cost to the nation of meeting its total energy requirements by alternative policies for the pattern or 'mix' of primary energy sources—coal, oil, natural gas or nuclear or hydro-electricity—over the fourteen-year period from end 1966 to end 1980. It claimed to have taken in all measurable economic costs, social as well as commercial. Consideration was also given to balance of payments effects, and to supply risks, taken as the extra cost that might have to be borne if the planned supplies (or costs) of one or more primary fuels failed to materialize.

The selected policies were six in number and covered a range of alternatives from the greatest use of coal to the least, allowing for varying degrees of dependence on oil, gas, or nuclear electricity. The range was wide enough to show the effect of moving in one direction or another, without going outside the limits of practical possibility. The policies were defined in terms of the combined amounts of various forms of energy, given for the key years 1970, 1975 and 1980. The first and most obvious feature to emerge, it was alleged, was that the total cost of supplying the nation's

[1] Letter in *The Times Business News* of 9 August 1968.

energy over the period from 1967 to 1980 would not be very different in percentage terms whichever of the alternative policies was adopted. The main reason for this somewhat surprising conclusion was the large area of ground which was common to all the policies—described as the facilities which were already in place at the end of 1966 and which must continue to be used, and the extra facilities which must be created in particular fuel industries and could not be switched from one industry to another within the limits of practical policy-making. This left only a restricted area for variations in energy policy. So, against the background of the country's total energy supply, the measurable cost differences were proportionately not very great. The cost differences were still quite large in money terms because the total costs discounted were over £19,000 million. But they had to be kept in perspective, and due attention given to the non-measurable social and other factors in the ultimate choice of policy.

One of the six policies came out easily the best. This was a middle-of-the-road mix of energy forms, which took advantage of the economics of the different forms in the output ranges where they were most favourable. For instance, it used the potential for higher productivity in the coal industry, as well as the most specialized attributes of oil and, still more, natural gas in the sectors where such attributes were worth their full economic price; and it allowed for an orderly progression in nuclear power. It avoided what were described as the very heavy capital costs of the largest programmes for North Sea gas, since it did not look as if a North Sea gas programme could justify itself beyond a scale at which it covered all the needs of premium-price markets and a limited amount of bulk sales to industry. Against the total cost of the country's energy supply, the measurable cost differences between the policies were, it was claimed, very limited. Whichever policy was followed for the flexible part of the supply, it was essential that all parts should work at reasonable levels of efficiency. If plans for some of the energy industries were made entirely on calculations restricted to those industries, the effect would be to throw out estimates of national energy cost completely. This could happen, for instance, if an unplanned reduction in certain coal markets led to such a fast rundown in colliery manpower that the coal industry was unable to retain a balanced work force, and could not therefore reach its productivity targets. The report did not conclude that there should be a reversion to the *status quo ante* in the fuel supply pattern when coal dominated the market. It conceded that advantages of new fuels should be vigorously pursued, but it did conclude that, to arrive at the future pattern, account should be taken of the whole spectrum of cost effects in all the industries, as well as the wider social and economic effects in the country.

Among the alternative policies in the report, the middle-of-the-road approach was stated to come out best because it employed the various forms of energy in the output ranges where they had the best economics; it avoided raising either coal or gas supplies to levels at which they were competing with more economically produced fuels; it avoided wasting the special merits of oil and gas in the special market areas where they could justify higher prices; and it avoided extra outlays on North Sea development merely to push a refined form of fuel up a factory or power station chimney.

Looked at objectively, the Report's claim to impartiality is, of course, difficult to concede. Its endorsement would moreover have meant the almost indefinite postponement of the adoption of a cheap energy policy at a time when Britain's main industrial competitors were rapidly reducing their energy costs by pit closures, increasing reliance on oil and natural gas, and pushing forward steadily with the long-term development of nuclear power. Thus, the case for coal was based, essentially, upon the realization of a productivity figure, by 1975, of 70 to 75 cwt per manshift—a figure whose actual realization must be regarded as a highly dubious proposition. Nevertheless, the Report had very real merits in underlining the many different strands that go to weave the overall tapestry of energy policy and, in particular, the importance of accurately calculating the total real costs as well as the total potential benefits of any particular line of policy.

Not all European coal industries have followed the line taken by their Belgian, British and German counterparts. In Holland, the Dutch Government was one of the first to decide that coal, at least in the European context, had no future and that there would, consequently, be little to gain from maintaining high levels of protection which were costly both to the taxpayer and to the consumer. Once this decision had been taken, the Government drew up and authorized a large scale redevelopment scheme designed to prevent the planned programme of total pit closures from causing a sharp increase in unemployment. This amounted, in fact, to a long term programme involving substantial participation by the public as well as some financial assistance from the European Commission in Brussels (i.e. the executive organ of the Common Market organization), which was geared to the creation of substantial industrial complexes designed to provide jobs for a large number of workers and to act as dynamic centres for the economic development for the region as a whole.

The largest project, which is being assisted by the Dutch Government by means of a guaranteed State loan of some £12 million and a Community loan of £4 million, was for the building of a car factory near the former Maurits colliery, once the pride of the European coalmining industry, which was closed down in 1966. The well-known and highly successful Dutch car manufacturing firm, Van Doorne's Automobielfabriek N.V. (D.A.F. for short) plan to use the new plant to produce new popular and medium sized models and so extend their production range, previously limited to the small 'Daffodil' model.

The project, which began in May 1966, is being implemented over a number of years. The first section, containing the hydraulic presses, came into use in July 1967, with a labour force of 500 men. By the end of 1969, the factory was expected to be able to produce cars without bringing any parts over from the main works at Eindhoven. By 1972 the project will have created some 6,000 jobs, at least half of which will be reserved for workers from the coal mining industry. However, it is significant that D.A.F. are not taking on anyone over the age of 45 and that, although they are obliged to take at least 50 per cent of their employees from the mining industry, this term is receiving a fairly broad definition as D.A.F. have found, predictably, that they have more use for semi-skilled workers from the technical industries which used to service the mines than from the mines themselves.

Other steps taken by the Dutch authorities have included the building of a medium-sized factory manufacturing carpets and floor coverings, with jobs for between 100 and 150 workers initially and some 250 workers in 1970; at a nearby small town with a sizeable commuter population, the High Authority of the European Coal and Steel Community granted a loan of some £150,000 towards the installation of a factory producing parts for heating and air conditioning equipment and sheet steel household articles—this factory was transferred with its entire staff from Maastricht and was expected to provide employment for between 100 and 150 redundant mine workers; at another key point in the former coal mining area is the site of a new modern brickworks using a special production technique—the firm concerned hoped to be able to increase its labour force from 100 to 140, mostly through taking on redundant mineworkers. Finally, in 1967 the High Authority of the E.C.S.C. agreed to a further loan of some £850,000 to assist in the installation of four enterprises for the manufacture of confectionery, mineral wool, electricity cable and plastic goods—the number of jobs to be created in this way was estimated at over 1,000, of which 900 were regarded as suitable for former miners.

In this connection, it is interesting to note that between 1954 and 1967 the High Authority of the European Coal and Steel Community paid out in re-adaption aids, grants and loans over £30 million in projects affecting some 275,000 workers in the coal, steel and iron ore industries of the Common Market countries.

The measures proposed in Holland and their subsequent success are undoubtedly very impressive. It would, however, be argued quite reasonably that the number of mines and men involved were small, and that the region in question, although not industrialized, was geographically well situated to encourage firms to take advantage of the low-interest loans offered to them by the Dutch Government and the European institutions. To see the problem and its proposed solution on a grander scale, we must turn to France, where the Government in 1968 decided, with characteristic Gaullist ruthlessness, to cut back the size of the coal industry from nearly 50 million tons in 1967 to 25 million tons or less by 1975.

The measures taken in France fall into two main categories. First, those designed to encourage miners to leave the industry, and, secondly, those geared to attract new industries to the coal mining areas in order to provide alternative employment for redundant workers. Steps taken within the first broad category have included enforced retirement of miners between 50 and 55 years of age with full pension rights, and for younger men payment of a reconversion bonus up to a maximum of one year's salary, a guaranteed make up for a year for any loss of salary incurred as a result of moving out of the coal industry, and the organization of special courses to provide training for alternative employment. The view of the French authorities is that money spent in this way is not only worth while but rich in rewards. French studies into both the costs of reconversion and the financial return upon public funds in this way are probably more detailed and sophisticated than any yet made in other countries. Basically the cost of retraining one miner for work in another industry is put at Fr. F. 32,000. The gross gain arising in the economy from this retraining and absorption into new fields of activity, on the other hand, is estimated, on the basis of an average further fifteen years of active work at between Fr. F. 50,000 and 150,000: a net gain, in other words, of not less than Fr. F. 18,000 when seen in the least optimistic light and as much as Fr. F. 118,000 under the most optimistic conditions. In these circumstances the incentives, from a national point of view, to persuade young miners to leave the coal industry as quickly as possible are clearly very great. This lesson has been taken to heart: steps have been taken to implement it and miners under 30 years of age are given extremely strong inducements to leave.

Measures taken under the second category—which we may fairly describe as a full blown regional reconversion policy—fall within the competence of a national commission, entitled 'La Commission pour l'aménagement du territoire', whose task it is to put forward proposals for regional industrial development and to co-ordinate the activities of local authorities, government and industry in this respect. While this commission has wide ranging national responsibilities, the French Government has imposed upon the Charbonnages de France the task of actively participating in the process of reconversion in the coalfields under its jurisdiction. While such a policy may smack at times of rubbing salt into open wounds, it is of course completely consistent with the conception that nationalized industries must operate, or be operated, in the best national interests. The measures taken to this end by the Charbonnages de France have consisted essentially in the creation of a wholly owned subsidiary company, owned and staffed by the Charbonnages under the name of SOFIREM, with a capital of Fr. F. 40 million, whose sole purpose is to assist in attracting new industries to areas where pit closures are planned or projected. SOFIREM may grant loans to, or take a minority holding in, these new companies. It is, however, a condition of these loans that there shall be no strings attached. Thus, the Charbonnages may not even take any special measures, such as particularly attractive but not generalized (and, therefore, discriminating) price rebates, to entice these companies to use coal in preference to other fuels. Finally, there must be ample provision for these firms to buy back the Charbonnages' minority holding in them as soon as they are in a position to do so.

The measures taken by the French authorities are particularly dynamic and thorough. Even so, given the size and magnitude of the problem, serious difficulties remain. For instance, in the Nord/Pas de Calais coalfield, where the geological conditions for mining are particularly unfavourable, the pit closure programme resulted in a net emigration from the region between 1962 and 1968 of nearly 100,000 people, despite the fact that this is, relatively, an underpopulated area and one where both the central Government and the local authorities have been actively pursuing an ambitious regional development policy. While a number of leading French industrialists have been prevailed upon in the last year or two to undertake to establish factories in the region—these include Renault, France-Couleur (manufacturers of television tubes), Simca and others—several years must necessarily elapse before this can be translated into fact; in the meanwhile, the Government has enforced an accelerated pit closure programme which has raised deep anxiety throughout the mining population in the area.

We said at the beginning that an evaluation of the losses and gains of a reconversion policy is not a simple matter. In terms of pure economics, there is only one solution: to close any pits that cannot compete with oil or nuclear power. Even the human or social problems can be resolved—at a price—through a far-sighted and generous policy of reconversion, resettlement and retraining, provision for optional anticipated retirement and acceptance, where necessary, of a considerable loss of useful manpower. While such a policy would cost dear, it is possible to argue that its medium or long term economic benefits would outweigh the high short term costs involved. The great question mark that arises, particularly in the case of the United Kingdom, from the potential effect of the closure of the coal industry upon regional development, is how the coal mines in areas like South Wales, Scotland and the north-east of England, can be replaced by alternative and at the same time economically viable industrial units. An unfortunate probability is that large tracts of an already over-populated land would become, first, economic and, later, human deserts, bearing the scars of their industrial heritage in their stultifying and newly found role as mausolea of technological obsolescence, with a blind disregard for the penalties that will have, one day, to be paid for the precipitate acceptance of a policy that cannot but accentuate the existing gross regional imbalances.

SECTION IV

THE DEMAND AND SUPPLY OF ENERGY

The rising demand for energy

As we have seen, contemporary civilization is based on large and regular supplies of power. Demand for energy is growing at an extremely rapid rate brought about by a combination of population growth, increasing general mechanization and industrialization, increasing demand for convenience and comfort in the home and, finally, new uses of energy.

Total world consumption of primary energy increased nearly threefold between 1929 and 1962. Certain types of fuel, like oil and natural gas, have shown even higher rates of growth. During the same time there has been a general trend towards transformation of primary into secondary forms of energy, particularly in the more highly industrialized parts of the world. This has been particularly true for the transformation of primary energy into electricity, demand for which has grown at a particularly impressive rate and which also has a number of specific uses.

Average energy consumption per head of population varies widely from one country to another, as we saw in Section I, and is dependent upon climatic conditions as well as on the prevailing degree of industrialization. The 1966 Report by the Energy Commission of the Organization for Economic Co-operation and Development (O.E.C.D.) on 'Energy Policy, problems and objectives' gave the following regional examples:

Consumption per head of population in 1962 in tons of coal equivalent

O.E.C.D. North America	8.6
O.E.C.D. Europe	2.8
Japan	1.6
O.E.C.D. area as a whole	4.4
U.S.S.R. and Eastern Europe	3.0
Communist Asia	0.6
Middle East and Africa	0.4
Caribbean and South America	0.7
Oceania	3.0
Far East (excluding Japan and Communist China)	0.1
World	1.4

But, as the O.E.C.D. report rightly commented, there are very big variations even within the European area of the organization, with figures ranging from less than 0·5 to about 3·5 tons a year. At the same time consumption in North America, although very much higher was still showing no signs of levelling off.

In trying to assess the rate of growth in world demand for energy, we have to take many different areas into account. In the Common Market countries for example, many detailed studies have been made, analysing almost every conceivable market development as far ahead as 1980, as the European Commission has struggled with its unhappy task of formulating a common, or at least co-ordinated energy policy for all six Member States. The O.E.C.D. has also carried out a number of perhaps less exhaustive but nonetheless extremely detailed surveys up to 1980. For other parts of the world, and particularly for the Communist bloc countries, there is much less information available. Indeed, many countries, particularly those which normally fall into the category of under-developed nations, make no detailed forward energy forecasts, and are content to rely on rudimentary, rule-of-thumb calculations.

We saw earlier on some of the difficulties that are involved in elaborating reliable energy forecasts. As the 1966 O.E.C.D. energy report stated: 'Forecasts of total energy demand can be made either in total, based on the general rate of expected economic growth or other global indicators, or on a separate examination of the main categories of consumption with a general assumption about the rate of economic growth.' For the second method of approach, energy demand was divided into three main categories of consumers: industry, transport and a 'domestic and miscellaneous sector' which in fact included consumption by commerce, agriculture and public and local administrations. At this point it must once again be emphasized that energy demand in each of these three categories is affected not only by the growth rate in economic activity, but also by the degree of industrialization and climatic conditions. Some heavy industries, such as the iron and steel industry, are, as we have seen, particularly large consumers of energy (in the European Community, for example, energy consumption in the iron and steel industry is equal to about 45 per cent of the total energy demand for all manufacturing industries taken together).

The rate of growth in demand for fuel for commercial transport purposes is intimately linked to overall economic growth. The same applies to private transport since the standard of living of any given country is a factor of the overall economic situation. Domestic heating and especially home heating standards constitute particularly good examples of this rule. The basic original need is for heating and cooking but, as the standard of living rises, so does the demand for space and water heating, for better lighting, for air conditioning, refrigerators and other electrical and labour saving appliances. Parallel with this comes a change in the type of fuel that is used, i.e. a switch from primitive fuels such as wood or charcoal, peat or dried dung for some of the under-developed countries, to coal, oil or natural gas. Later these tend in their turn to give way to electricity which, although more expensive, is a much more convenient form of energy.

In this situation of rising energy demand the principal factor making for a reduction in energy consumption is the continual improvement in fuel burning efficiency, either through the development of new or improved appliances or equipment, or the substitution of new forms of energy for old. This is an important consideration since, without these developments, industrial costs would have been substantially higher and this, in turn, could well have contributed towards bringing about a lower level of industrial activity. But

> while it is true generally that energy demand is dependent on economic activity, the relationship between the growth of the former and that of gross national product as a measure of the latter, it is difficult to interpret as it is affected by structural changes within and between the different sectors of economic activity. Differences are large between countries, and for any single country this ratio (i.e. the relationship between rising energy demand and the rate of increase in economic growth) may change over time. In Japan, the rate of growth of energy demand in post-war years has exceeded that of G.N.P., as was often observed for other countries in the course of rapid industrialization. In the European region of the O.E.C.D. as a whole, the ratio of these growth rates has been less than unity. In the North American area energy demand likewise rose somewhat slower than G.N.P. in recent years.[1]

There are also the effects of technical innovation and Government interventions to take into account. In the same way as, in the past, the development of the steam and internal combustion engines had a revolutionary effect upon energy demand, so in our own times the demand for fuel in the transport sector has been subjected to fundamental alterations in volume as well as in pattern by the introduction of the jet engine and the new fuels required in space exploration, while the prospects opened up by nuclear power are boundless.

Government policies affect the development of the overall energy situation in three ways: by their direct or indirect control over fuel prices; by their influence on the overall rhythm of economic growth; and by their

[1] Energy Policy: problems and objectives, O.E.C.D., Paris, 1966, p. 21.

general social and economic policy. The 1966 O.E.C.D. Report stated that, at any given time, the pattern of energy demands 'is formed by the specific demand for certain fuels for which there is no technical alternative, by price competition between fuels in so far as it determines the cost for the consumer, and by consumer preferences based on considerations of convenience or related factors'. We do not have to look far to find examples of fuels meeting the specific types of demand, i.e. automotive fuels for road and air transport, petroleum products for non-fuel uses such as lubricants or chemical feedstocks, or coking coals for the iron and steel industry (although in this last case there are, as we have seen, new processes being developed, such as the injection of oxygen in various new steel-making processes and the injection of liquid and gaseous hydrocarbons through the tuyeres of the blast furnaces, which are eroding the hitherto monopolistic position of coking coal). In total, it was found that, in 1964, forms of specific fuel consumption accounted for about one-third of all energy demand in the O.E.C.D. area.

In the remainder of the energy market where one fuel is readily interchangeable with another, the most important single factor influencing the consumer's choice or preference is undoubtedly, and understandably, price, particularly in the case of industries that are large energy consumers. There may, however, be certain cases, e.g. the chemical industry or the glass-blowing industry, where ease, convenience and flexibility of handling may be regarded as being of almost equivalent importance. In some respects, this latter comment applies even more forcibly to the domestic sector than it does to the industrial, since a householder, particularly in the richer and more advanced industrial countries, is often prepared to pay handsomely for additional convenience. The 1966 O.E.C.D. Report stated, notably, in this respect that

> in industry, costs are more important in view of the need to remain competitive; consequently the costs of handling, storing and firing various fuels are taken into account when choosing between different fuels. There are, however, instances where a fuel has individual qualities—e.g. that it is immediately and finely adjustable or that it produces a sulphur-free flame—which makes it preferable for particular processes or for the manufacture of particular products. Also, the requirements of modern processes have accentuated the industrial demand for refined fuels, as has also to some extent the demand for clean air.[1]

Most of what we have said so far may be held to apply with almost equal force to all the industrialized regions of the world. We must now, however, begin to break down the overall global energy picture into a

[1] *Ibid.*, p. 22.

number of constituent parts. If we take first of all the O.E.C.D. area, we have at our disposal the detailed forecasts made in 1966 by the Governments of the countries concerned up to 1980. In their calculations, the forecasters made certain assumptions

> about the possible rate of economic growth; in order to take account of the varying effects of the different elements in the economy, projections for the various sectors of demand have been made; estimates have been made of specific demand for coking coal and for certain petroleum products; particular attention has been paid to the probable growth in requirements for electricity generation, taking account of the likely role of nuclear power. Estimates of total requirements related to the assumed level of economic growth have also been made. We also consider later possible future trends in the final prices of the different fuels.
>
> All the estimates have assumed that there would be regular economic development with no major crises; that there will be regular development in the technology of energy production, transport and use; that there will be continuing competition between the different forms of energy, subject to government intervention which is assumed to continue on the same lines as at present.[1]

Besides global estimates for the O.E.C.D. area as a whole, sector forecasts were also prepared for the three main geographically component regions, i.e. North America, Western Europe and Japan.

North America

In North America, energy consumption between 1929 (the last so-called 'ordinary' year before the depression of the 1930s) and 1964 rose from 586 million to 1,341 million tons of oil equivalent. The rate of increase had, however, been far from uniform: thus, an extremely low rate of increase during the years between 1929 and 1936 had been followed by a rise of no less than 176 per cent between 1937 and 1964. During the first half of this latter period, energy demand increased at a rate of 4·5 per cent, and this had fallen between 1950 and 1964 to approximately 3 per cent. Even so, the average *per capita* consumption of energy was still nearly three times higher than the average figure for Western Europe. As for the future, an annual rate of increase between 1964 and 1980 of 3·7 per cent was forecast. By far the biggest rate of increase was expected in the electricity generating sector which was expected to grow by two and a half times during this period and to account for no less than 35 per cent of the increase in primary fuel requirements. By 1980, it was envisaged that the electricity generating sector's primary energy requirements would represent some 30 per cent of total fuel requirements compared with 21 per cent in 1964.

In addition to growth estimates in individual sectors the Energy Com-

[1] *Ibid.*

mission also made a global estimate based on two main assumptions: that there would in fact be a straightforward 4 per cent annual rate of increase in Gross National Product and the simple projection up to 1980 of the steadily declining relationship between the amount of energy required per additional unit of Gross National Production that had been a feature of the last thirty years. The results obtained from these two different methods of approach differed, on average, by a mere fraction in 1970 and by only about 3 per cent in 1980. The Energy Commission also tried yet a third approach which consisted of projecting *per capita* consumption trends; this method produced somewhat lower figures. Finally, however, the Energy Commission turned back to the individual projection on the grounds that these returns, more than any others, could be relied upon to make adequate allowances for different rates of development in different sectors of the economy.

Table 9. *O.E.C.D. North America estimates of energy requirements according to the sectors of demand*

	Million tons of oil equivalent		
	1964	1970	1980
Final consumption in:			
Iron and steel / Other industry / Energy sector	475	570	750
Transport (including bunkers)	284	353	511
Domestic and miscellaneous	342	427	619
Total final consumption	1,101	1,350	1,880
Transformation losses and non-energy products	240	330	520
Total primary energy requirements	1,341	1,680	2,400
Of which for electricity generation	282	387	657

The European area

In the O.E.C.D. European area, total final consumption between 1950 and 1964 grew by 3·9 per cent a year. Forward predictions up to 1980 suggested a slight slackening in the rate of increase to about 3·7 per cent a year. It was suggested that while the fuel requirements of the transport sector might grow at a faster rate than those of other sectors, the final breakdown of energy demand, like that in the United States, would not reveal any major departures from the existing pattern.

The estimates of primary fuel requirements for electricity generation had, it was stated, been derived from a careful analysis of possible require-

[1] *Ibid.*, p. 26.

ments in the major consumer sectors. It is, of course, well known that electricity has been rapidly increasing its share of the market. The O.E.C.D. Energy Commission, for instance, summed it up neatly by saying that 'for some purposes it is essential, for others it is highly competitive, while for still others its convenience and flexibility can provide ready markets for it. 'The accuracy and appositeness of this statement are

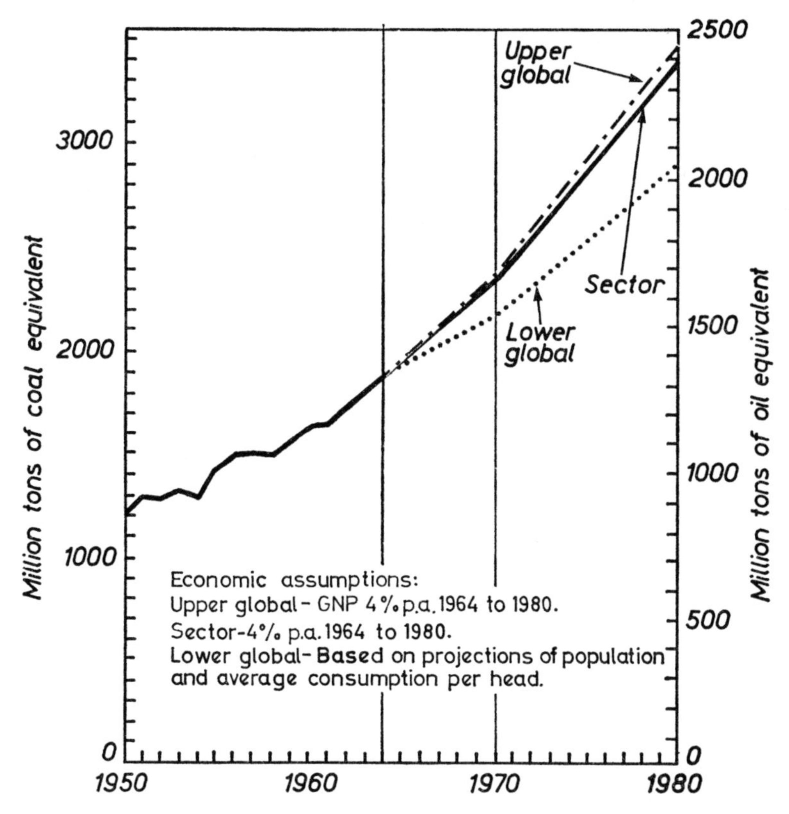

Figure 5. North America: estimates of total primary energy requirements 1970 and 1980
Source: 1966 O.E.C.D. Report. 'Energy Policy problems and objectives', p. 27

certainly no less true, and possibly a good deal more so, for the intervening period up to 1980. For the ten years up to 1964 the annual rate of increase had been a little less than 8 per cent; for the next fifteen years the Energy Commission's sector analyses suggested a rate of growth of approximately 7 per cent—although there are a number of experts who regard this figure as excessively conservative. Even on this basis, however, total gross output of electricity would increase from 756 terawatt hours (TWh) in 1964 to 2,130 TWh by 1980. The biggest increase in percentage terms is expected to take place in the domestic sector. In their efforts to compute the fuel requirements of the electricity generating sector, the Energy Commission assumed that there would be considerable further progress in fuel burning efficiency as a result of continuing technical developments and the introduction of larger power stations; these were expected to bring about an increase in average efficiency from 30 per cent in 1964 to 35 per cent in 1970 and 38 per cent by 1980. Total demand for primary energy by 1980 was nevertheless expected to be three times as great as in 1960. 'The significance of these figures', the O.E.C.D. report duly commented,

> should not be ignored. They imply that, between 1964 and 1980, gross consumption of electricity, rising from 65 million tons of oil equivalent (756 TWh) to 183 million tons (2,130 TWh), will nearly double its share of total final energy consumption and account for approximately 20 per cent towards the end of the period. Requirements of primary energy for electricity generation would account for nearly half the increase in the total primary energy demand between 1964 and 1980 and for 32 per cent of the total used in the latter year. It is true that these estimates are related to a high level of sustained economic growth throughout the period; but even if the high level is not attained, it seems likely that total fuel consumption would be more significantly affected than would electricity consumption.[1]

Specific demand for petroleum products was expected to grow even faster in Europe than in North America (for although there was still no sign of anything remotely approaching saturation in this area, the demand curve in the United States and Canada had nonetheless begun in the last few years to show some signs of levelling out); as a result oil's specific share of the European primary energy market was expected to increase from 17 to 21 per cent. The main contributory reason for this increase was the growing popularity of motor transport and civil and private aviation. In the railway sector big increases in efficiency were anticipated and these would, it was believed, more than offset any increase in energy demand arising from possible increases in rail traffic.

The Energy Commission made a check on its sector forecasts by means of two global estimates. The first of these assumed an annual increase of 4·5 per cent; while there are manifest differences from one country to another, it was nevertheless felt that this figure was a realistic representation of the upper end of the likely range. The Energy Commission's second global estimate assumed an annual growth rate in Gross National Product of 4 per cent over the period under review. Basing themselves on the fact that the relationship between the average rate of growth in energy

[1] *Ibid.*, p. 32.

demand and the increase in gross national production was broadly equal to parity in the 1950s and to about 0·8 during the fifteen year period from 1950 to 1964, the Energy Commission assumed a slightly rising trend in their first global estimate and a constant figure of 0·8 in the second; this resulted in ranges of global demand figures of 880 to 925 million tons of oil equivalent in 1970 and 1,200 to 1,390 million tons in 1980.

In the event, the estimates based on the sector approach were about 4 per cent higher than the figures obtained from the global estimates. The Energy Commission accordingly opted, as in the case of North America, for the sector approach. Explaining their decision, the forecasters said that

> global estimates are useful for giving broad indications of magnitude, but the more detailed sector analysis should be more reliable. Secondly for the purpose of this report we want to be most careful not to underestimate requirements, and therefore we prefer to adopt the sector estimate as an upper limit. It may be noted that energy consumption per head may rise from 2·2 tons of oil equivalent in 1964 to 3·6 per head in 1980.

Table 10. *O.E.C.D. Europe: estimates of energy requirements according to the main sectors of demand*[1]

	Million tons of oil equivalent		
	1964	1970	1980
Final consumption in:			
Iron and steel	66	76	88
Other industry	158	196	282
Transport (including bunkers)	119	161	244
Domestic and miscellaneous	193	240	358
Energy sector	65	77	108
Total final consumption	601	750	1,080
Transformation losses and non-energy products	150	210	360
Total primary energy requirements	751	960	1,440
Of which for electricity generation	185	259	460

Japan

In Japan, there has, as we have already seen, been a particularly dramatic growth in energy demand over the last ten to fifteen years. A glance at the statistical tables shows that energy consumption in 1964 was four and a half times higher than it was in 1929; this, however, does not reflect the startling fact that the actual energy consumption figure for 1950 was

[1] *Ibid.*, p. 32.

virtually identical with that for 1929. In other words the whole of this remarkable increase has taken place during the postwar boom period which began in 1950 and is still showing no signs of slackening off. In fact, even a comparatively superficial examination of the Japanese energy figures will reveal that the rise in energy consumption has been closely linked to the country's industrial explosion: the general industrial sector

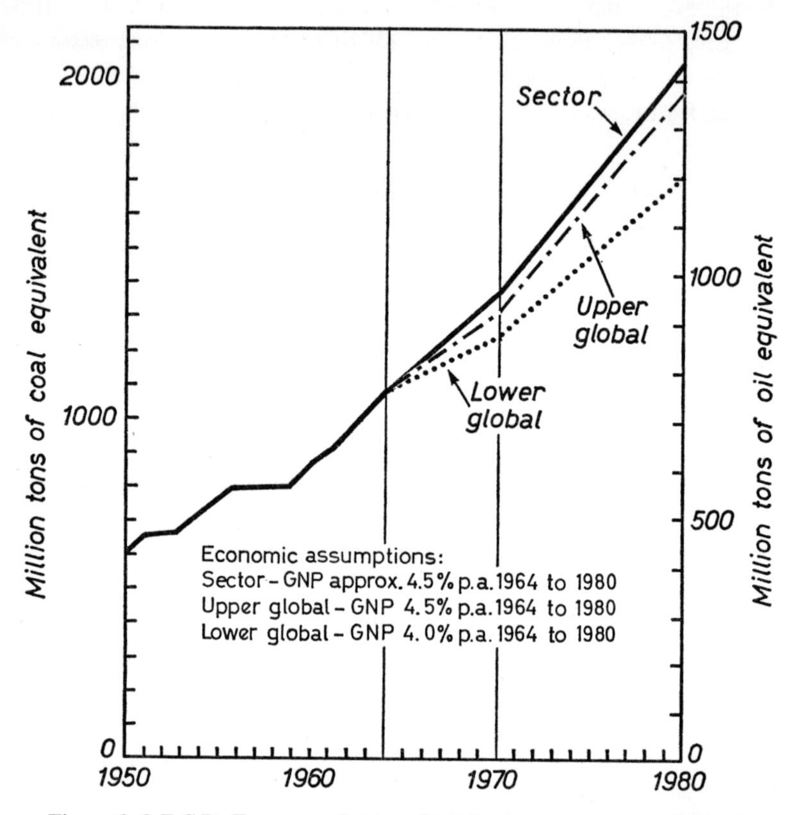

Figure 6. O.E.C.D. Europe: estimates of total primary energy requirements 1970 and 1980
Source: 1966 O.E.C.D. Report. 'Energy Policy problems and objectives', p. 33

accounts for the remarkably high figure of 50 per cent of total national energy consumption and has, indeed, slightly increased its share since 1950. For the rest, transport requirements represent about 23 per cent and the domestic and miscellaneous sector about 19 per cent of total national energy consumption (it is interesting to note by way of contrast that the general industrial sector represents about 40 per cent of final consumption of energy in the European O.E.C.D. area and less than

30 per cent in the North American area; in both areas these proportions are, moreover, tending to decline).

In making these forward projections of energy demand the Commission preferred to use forecasts based on the growth rates in gross national production over the past few years and the main trends that emerged from them, rather than the official figures prepared by the Japanese Government. The result was an estimated rate of increase in energy demand of some 8 per cent a year (or, more precisely, of 8·3 per cent up to 1970 and 7·2 per cent from 1970 to 1980). While this is, of course, a remarkably high figure, nearly twice as high as the rates assumed for North America and Europe, it is, in fact, a little below the average annual rate of increase that was actually achieved between 1950 and 1964.

Table 11. *Japan: estimates of energy requirements according to the main sectors of demand*[1]

	Million tons of oil equivalent		
	1964	1970	1980
Final consumption in:			
Iron and steel	17·3	25	31
Other industry	33·7	61	139
Transport (including bunkers)	23·2	41	88
Domestic and miscellaneous	18·8	25	36
Energy sector	8·3	10	16
Total final consumption	101·3	161	310
Transformation losses and non-energy products	33·0	77	154
Total primary energy requirements	134·3	238	464
Of which for electricity generation	39·8	74	152

Japan provides a number of remarkable and fascinating changes in energy demand patterns from those which we have come to associate almost automatically with advanced industrial societies of the West. Thus, while the energy requirements of the transport sector are likely to grow rapidly and to increase their share of total energy demand from 23 per cent in 1964 to 28 per cent in 1980, the part of total demand accounted for by general industry is expected to remain at more than 50 per cent while that of the domestic and miscellaneous sector will almost certainly fall. One outstanding feature that is common to all three regions, however, is the continuing very rapid growth in demand for electricity.

[1] *Ibid.*, p. 38.

As in the case of North America and Europe, the Energy Commission also made projections of forward energy demand in Japan on a global basis, using the same assumptions about overall economic growth as in the sector approach. The ratio between rising gross national production and rising energy demand was considered to be, at the present time, at parity and was expected to remain fairly constant at about this figure for the period up to 1980. The figures finally used by the Energy Commission were once again, however, those obtained by means of the sector approach.

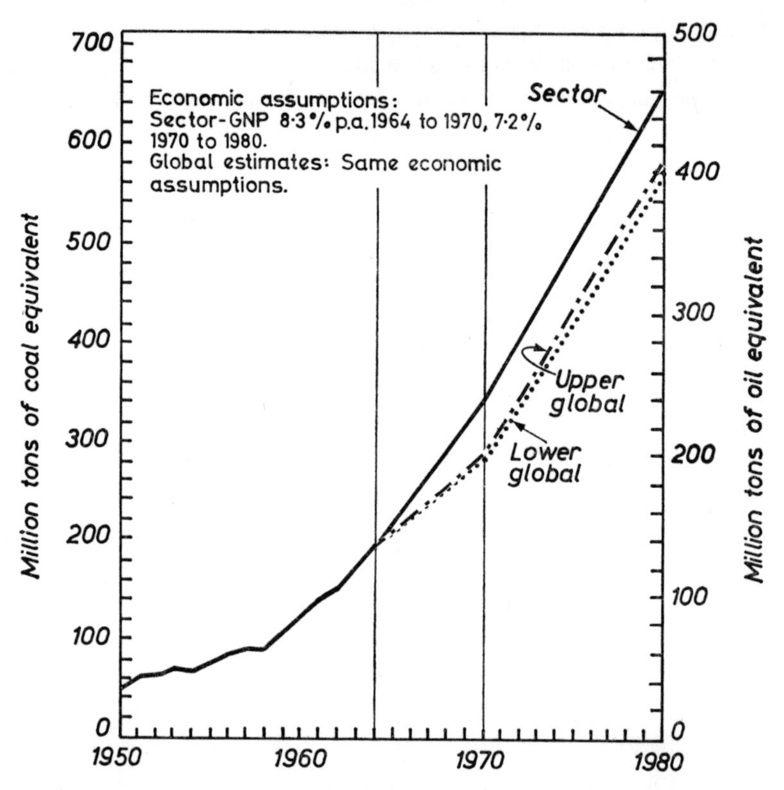

Figure 7. Japan: estimates of total primary energy requirements 1970 and 1980

Source: 1966 O.E.C.D. Report. 'Energy Policy: problems and objectives', p. 39

Conclusion

These figures, as the Energy Commission noted with an unusual degree of exuberance, showed 'a highly staggering rate of increase'. It should not, however, be overlooked that even as recently as 1964 the *per capita* level of energy consumption in Japan, at 1·4 tons of oil equivalent, was much lower than in Europe with 2·2 tons, or North America with 6·3 tons of oil equivalent. By 1980 the *per capita* figure for Japan is expected to have reached between 3·6 and 4·1 tons, compared with an average European figure by that time of 3·6 tons (with the most advanced European industrial countries notching up figures of just over 5 tons) and about 8·7 tons of oil equivalent for North America.

Total energy demand in the whole of the O.E.C.D. area was accordingly expected to rise from 2,225 million tons of oil equivalent in 1964 to 2,880 million tons in 1970 and 4,300 million tons by 1980. These estimates, if proved correct, would mean therefore a near doubling of primary fuel requirements in the period under review, with an annual rate of increase of 4·2 per cent, compared with about 3·9 per cent in the years between 1950 and 1964. *Per capita* consumption would show an increase from three to five tons of oil equivalent. Although energy demand is expected to continue to increase in all the principal sectors, it is in the general industrial sector that the most massive increase—at least in absolute terms—is expected; in relative or percentage terms the transport sector seems certain to lead the poll. As a result, and as a function of developments in these two sectors, it is confidently predicted by most experts in the field of international energy forecasting that the demand for oil products especially will more than double between 1964 and 1980.

Before going on to consider demand trends in other parts of the world, it is both useful and desirable to examine with some care the tremendously detailed forecasts that have been prepared by the Commission of the European Communities (more often referred to as the Common Market Commission) for the six Common Market countries (i.e. Belgium, France, West Germany, Italy, Luxembourg and the Netherlands). Prepared with meticulous care these forecasts are the result of some of the most advanced methodological and forecasting techniques, which virtually no other countries or areas are in a position to match in terms of attention to detail and reasoned or calculated plausibility.

The forward demand situation in the Common Market countries

The Community's most recent forward estimates[1] for the period up to 1980 can be conveniently subdivided into two sections: the short term period up to 1970 and the long term forecast for the following ten years. For the short term period the Community experts who, it must be emphasized, had free and detailed access to the competent Government departments in

[1] Nouvelles réflexions sur les perspectives à long terme de la Communauté européenne, Luxembourg, 1966.

the six Member countries, assumed an overall rate of growth in Gross National Product between 1965 and 1970 of 4·6 per cent: the individual rates being expected to be somewhat higher in France, Italy and and Holland and somewhat lower in the other three countries. These figures were of course no more than targets whose realization, it was admitted, depended upon intensive investment programmes and optimum utilization of all available manpower as well as the approximate stability of the general level of prices and the equilibrium of the balance of payments. It followed therefore that the adoption of measures likely to make more manpower available and to reduce production costs was a vital prerequisite for the achievement of the desired rate of expansion and the corresponding improvement in the standard of living.

In the course of a long, exhaustive but extremely careful survey of the forward position of the main consumer sectors, the Community experts forecast, in particular, a continuing high rate of growth in demand for electricity, at a rate slightly in excess of doubling every ten years. Essential characteristics in this sector were stated to be the extremely modest rate of increase in production that can be expected from hydro-stations or plants burning lignite or blast furnace gas, estimated at no more than 30 TWh out of an additional 190 TWh during the period. With the contribution from nuclear plants not expected to exceed 25 TWh, this left 125 TWh to be provided from conventional sources, notably oil and natural gas. In the steel sector, total production was expected to rise at a modest rate from 85 million tons in 1965 to 95 million tons by 1970; at the same time continuing technical advances in the improvement of blast furnace efficiency were expected to result in a further decline in the level of specific requirements of coke for the production of one ton of steel from an average figure of 700 kg in 1965 to some 600 kg by 1970. In the general industrial sector there were two main points of interest: first, a slowing down in the rate of progress achieved in improving fuel utilization, since the wave of technical innovations which had been such a feature of the period from 1950 to 1960 had spent a good deal of its vital initial force. As a result, the anticipated rate of increase in industrial production of over 5 per cent a year was expected to give rise to an increase in energy demand of not less than 4·5 per cent—a somewhat high ratio for the Community countries. The second feature of developments in this sector was the continuing switch from coal to other fuels. Even on a basis of broadly equivalent prices, the advantages of using hydrocarbon fuels were so substantial that the use of coal in this sector seemed destined to vanish almost completely. In the transport sector, similarly, a big the same annual rate as for the short term period up to 1970. In the electricity generating sector, a continuing high level of demand was expected as a result of further increases in the use of electrical household equipment and general industrial requirements, with the result that the growth rate of the last fifteen years will, in all probability, be maintained. More precisely, the demand for electric current was expected to increase during the fifteen-year period from 1965 to 1980 by a factor of 2·8 in general industry, by a factor of 4 in the domestic sector and by a factor of ·9 overall. As a result, total demand for electricity was expected to soar rom 416 TWh in 1965 to 1,200 TWh in 1980. A development of this rder of magnitude could, however, be achieved, it was emphasized, only if ne relative fall in electricity prices in relation to the general level of prices ere to continue. On the production side, it was felt that the possibilities f further technical improvement at conventional thermal power stations ere limited and that the future lay essentially in nuclear power stations. n the steel industry front, steel output was expected to increase com- aratively slowly in the 1970s: at the same time, the industry would, it was lt, still be very far by the beginning of the next decade from having hausted all possibilities of further reductions in specific coke oven nsumption by means of using richer or enriched ores, greater use of ger ovens and fuel injection. As modernization is essential in the ghly competitive steel market, it was believed that these technical portunities would be exploited to the full, leading up to a further crease of up to 30 per cent in specific coke consumption to some 480– kg per ton of steel or less. In the case of general industry, it was nsidered unlikely that savings arising from further improvements in l utilization would exceed 1·5 per cent of total annual expenditure energy supplies, compared with 3·1 per cent a year in the period from o–60. In view of this, and assuming an annual industrial growth rate of r cent, it was likely that energy consumption by industry would show increase of some 4 per cent a year to reach a total by 1980 of over million tons of coal equivalent. While oil would remain the principal ce of energy for this sector, it was believed that consumption of ral gas would rise rapidly and might account for as much as 30 per cent is market by the end of the period. In the transport sector the antici- d expansion of private motoring and all forms of commercial transport expected to provide a continuing stimulant to demand for motor fuels lead to a doubling of total energy consumption to some 165 million of coal equivalent. Turning lastly, once again, to the domestic r, the Energy Commission recognized that this was the most difficult

increase in demand for oil was predicted. Finally, in the
the signs were that higher living standards were encouragi
extend home heating facilities at a steadily acceleratin
showing no signs of approaching saturation. It was felt
energy could increase by up to 20 per cent in the five-y
1970. Consumption of solid fuel, although reasonably s
had since been falling off steadily, and it seemed probabl
would make further rapid inroads into its market.
energy consumption in the Community by 1970 was
743 million tons of coal equivalent or 25 per cent m

Turning to the longer term prospects up to 1980, the C
readily admitted that their estimates were, in this case
speculative nature and based on information that w
little more than fragmentary. The Community ex
advanced the following reasons for their belief that
consequently, gross national production would co
rapid rate of increase: first, individual countries w
research efforts as they became increasingly aware of
new technologies; this in turn, would facilitate the ad
niques and procedures and a more rapid comme
introduction of new products. Second, even within
technological expertise, there was an enormous pot
productivity. This was borne out by the substanti
ductivity and effectiveness between different branc
different enterprises within one industrial field. In
was still entirely possible and, indeed, desirable to obt
productivity from the simple expedient of withdrawi
productivity occupations and directing them tow
occupations or industries that gave a high rate of
Finally, the abolition within the frontiers of the
obstacles to the free circulation of goods, capital
gressive application of common policies should
a substantial improvement in the effectiveness of th
and, in the process, make a substantial contributior
technological gap between Europe and the United

On the basis of these considerations and allow
increase in numbers of the working populatio
Germany and limited to between 0·5 and 1 per c
five countries, the Community experts finally took
for the increase in gross national production du

c
v
C
p
fc
e:
c
la
hi
o
de
5c
co
fu
on
19
5
an
20
sou
nat
of
pat
was
and
ton
sect

one to forecast with any real degree of precision. *Per capita* energy consumption in this sector in 1965 at 760 kg was well below the comparable American figure of 1,870 kg; there was, moreover, still no sign of any slackening in the rate of increase in demand so that an overall figure of 220 million tons of coal equivalent by 1980 for this sector seemed quite realistic. The market shares of the various fuels were likely to depend at least as much on convenience as on price and were expected to result in a major slump in consumption of solid fuels and a big increase, possibly as much as 40 per cent of the market, in demand for natural gas. In total, the energy requirements of the Common Market countries in 1980 were expected to reach a level of 1,130 million tons of coal equivalent, an increase of some 90 per cent over the 1965 figure of 596 million tons.

The forward demand situation in the United Kingdom

The United Kingdom's official forward energy estimates up to 1975 were published by the British Government in the form of a White Paper[1] in November 1967. This showed that total energy consumption in Britain was expected to rise from 297·7 million tons of coal equivalent in 1966 to 310 million tons in 1970 and 350 million tons by 1975. These figures were based on the assumption that there would be an annual rate of increase in economic growth of about 3 per cent. Although the White Paper did not give a global figure for 1980, this was widely reported in the British Press at the time to have been put, unofficially, at 400 million tons of coal equivalent.

While there was no attempt to lay down, categorically, market shares for the different types of fuel, the general tone of the White Paper quite clearly reflected a good deal of official pessimism about the ability of coal to compete with the other main sources of energy. As a result, energy demand for the period up to 1975 was expected to show the following pattern of development:

Table 12. *United Kingdom: forward estimated demands for energy*

(In million tons of coal equivalent)

	Actual 1966	White Paper Estimates 1970	White Paper Estimates 1975	Unofficial Forecast 1980
Coal	174·7	152	120	80
Oil	111·7	125	145	160
Nuclear and hydro-electricity	10·2	16	35	90
Natural gas	1·1	17	50	70
Total	297·7	310	350	400

[1] White Paper on Fuel Policy, Cmnd. 3438, H.M.S.O., London, 1967.

The forward demand situation in the Soviet Union and Eastern Europe

Information about the forward demand situation in the Soviet Union and Eastern European countries is sparse and generally rather unreliable. Energy demand in Eastern Europe, it has been estimated, will increase from its 1965 level of about 360 million tons of coal equivalent to some 700 million tons by 1980—equivalent to a rising annual demand of 5 per cent. In the case of the Soviet Union, the application of the historical rate of increase in energy demand of the last two decades to the period up to 1980 gives a total demand figure by that year of over 2,000 million tons—which would give the Soviet Union a demand figure for energy approximately equivalent to the entire European O.E.C.D. area. An even more ambitious estimate, however, was given by the Russian energy economist Melent'ev in 1962 when he forecast overall energy consumption figures for the Soviet Union of 1,371 million tons of coal equivalent in 1970 and 2,950 million tons by 1980; of this latter figure oil and natural gas would account for 1,850 million tons of coal equivalent and solid fuels for a further 850 million tons.[1]

The global forward demand situation

For most of the remaining parts of the world, the available information is either sketchy, partial or confusing. Most of the countries concerned do not prepare detailed forecasts but are content, as we saw in Section I, to rely on forward projections based on past trends. While techniques of this kind are clearly not very scientific or sophisticated, they are nonetheless about as near the mark as any other method that could be devised for a number of primitive or under-developed countries. The margins of error may, admittedly, be large, at least in relative terms, in some countries; but in the majority they are more likely than not to be reasonably accurate. Since, moreover, the effective absolute level of energy consumption in these countries is generally low, even a number of sizeable inaccuracies in a dozen or more countries in this category would have only a very slight marginal bearing on the overall situation.

Global world forward energy demand estimates up to 1980 were prepared both by the Energy Commission of O.E.C.D. and by the Commission of the European Communities. Their conclusions were broadly in line. The overall rate of increase in economic growth during this

[1] L. A. Melent'ev *et al.*, *The U.S.S.R. Fuel and Energy Balance*, Moscow and Leningrad, 1962. See also W. G. Jensen, *Energy in Europe, 1945–80*, Foulis, 1967, pp. 162–6.

period (excluding the Soviet Union and other Communist states) was estimated at between 4 and 5 per cent a year. Both reports agreed that the biggest uncertainty concerned the under-developed countries. The Community experts opted for an annual rate of increase in these countries of 6 per cent, although this seemed to many to be too optimistic and on the high side; even so, however, the countries in question would still have an extremely low *per capita* energy consumption figure by the end of the period under review. It was of course always possible that the sudden emergence of a dynamic growth rate in one or more areas could give rise to a dramatic change in the overall energy demand situation; there was, however, no indication so far that such a development was likely. Indeed, there was a good deal of evidence that pointed rather to a decline from the present rate of increase, subject to a few notable exceptions, such as Brazil. As far as the Communist countries were concerned, both the Community experts and the O.E.C.D. Energy Commission were prepared, in the absence of more reliable data, to work on the assumption that the annual average rate of increase both in economic growth and in energy demand would, as has been the case in the last few years, be slightly above the average level for non-Communist countries. These hypotheses led to the supposition that world energy demand would increase during the period up to 1980 by 4·7 per cent (compared with an annual growth rate of 5·1 per cent between 1953 and 1963). As a result, total world requirements of energy were expected to show an increase from 4,350 million tons in 1960 to 6,800 million tons in 1970 and 10,900 million tons by 1980 by the experts of the European Communities, or to 11,770 million tons of coal equivalent in the latter year by the O.E.C.D. Energy Commission. These totals, if they are achieved, will mean that in 1980 the energy requirements of the industrialized countries of the non-Communist countries (i.e. the United States and Canada, the Common Market, the United Kingdom and Japan) will still represent 50 per cent of the world's total energy consumption (of which the Common Market countries will represent 10 per cent and North America nearly 30 per cent).

The O.E.C.D. Energy Commission, commenting on its forecast figure of 11,770 million tons of coal equivalent by 1980, was clearly in an optimistic (perhaps excessively so) frame of mind, when it noted that

these figures suggest that while the rate of growth of energy demand in the O.E.C.D. Area over the years to 1980 will be about 4·3 per cent per annum, the increase in all other regions combined will be about 6 per cent compared with about 8 per cent per annum between 1950 and 1960. High growth rates of commercial energy consumption will be due in part to substitution of non-commercial forms, such as wood and peat. Also, the potential for industrial expansion in many of the non-O.E.C.D. countries is very great and, given favourable economic conditions, there is little reason why it should not take place. In any event, for the purposes of this report, it is preferable to over-estimate demand than to risk giving too comfortable a picture about the availability of supplies by underestimating.

On these prospects, therefore, the share of the world total consumed in the O.E.C.D. area would continue its fall (from 70 per cent in 1950 to 60 per cent in 1960) and be around 50 per cent by 1980. While *per capita* consumption would rise by some 65 per cent in the O.E.C.D. area between 1962 and 1980 to over 5 tons of oil equivalent, it would more than double in the rest of the world (0·5 to 1·1 tons of oil equivalent). Outside the O.E.C.D. area, only in Eastern Europe, U.S.S.R. and Oceania would *per capita* consumption be over 2 tons of oil equivalent and the average for the remaining regions—nearly three-quarters of the world's population—would still be only 0·5 tons of oil equivalent.[1]

Table 13. *Total world energy demand: 1960–70–80*

(In million tons of coal equivalent)

	Rate of increase (in per cent per year)		Total energy requirements			Share of total world requirements (%)	
	1953–63	1960–80	1960	1970	1980	1960	1980
Western Europe	+4·2	+4·2	845	1,300	1,920	19	18
North America	+3·0	+3·6	1,580	2,270	3,170	36	29
Japan	+8·6	+7·8	115	275	540	3	5
Latin America	+7·7	+6·8	155	295	555	4	5
Africa	+4·6	+4·4	75	110	180	2	2
Middle East	+9·5	+7·6	50	95	190	1	2
South and S.W. Asia	+7·9	+6·1	115	210	380	3	3
Polynesia	+4·1	+5·1	50	80	130	1	1
Communist countries	+8·3	+5·3	1,365	2,175	3,835	31	35
World	+5·1	+4·7	4,350	6,810	10,900	100	100

Energy availability

The overall energy availability situation can be assessed either by an examination of the supply and demand position of the various fuels or by splitting the world into a number of geographical units and doing a sum of their respective net import or export balances. This latter procedure was in fact the method of approach adopted by the Coal Producers of Western Europe in a defensively orientated propaganda booklet entitled *An Energy Policy for Western Europe*,[2] in which they divided the world into eleven regions and concluded, predictably, with the view that unless coal pro-

[1] O.E.C.D. Report. 'Energy Policy: problems and objectives', *op. cit.*, p. 42.

[2] Published by the Western European Coal Producers and the National Coal Board, March 1966.

duction could be maintained at about its existing level there would be a net shortage of energy well before the end of the century. This shortage was likely to come about in one of two ways. Energy consumption in the underdeveloped parts of the world would rise very slowly because of a basic failure to improve the general economic conditions in these countries. This in turn would then accentuate the difference in standards of living and wellbeing between rich and poor countries, which would inevitably lead to sharper political tensions with potential effects upon the availability of energy, particularly oil supplies from these areas, that were 'incalculable'. Alternatively, there would take place a rapid economic expansion of these countries which, while it would remove the root cause for political turmoil and popular discontent, would, on the other hand, generate such an increase in the demand for energy that their exporting capacity would once again be affected. In short, a classic example of the 'heads I win, tails you lose' argument.

The Coal Producers were right, of course, when they emphasized the fact that there were in the world of energy supply and demand three main types of region: those that are more or less self-sufficient; those that are largely dependent upon imports; those that have big energy surpluses. Among the countries or areas which can be described as belonging to the larger self-sufficient category are the United States, and the U.S.S.R. and Eastern Europe taken as a single unit. Among those with large export surpluses are the Middle East, North Africa, Nigeria and Venezuela. Among those, finally, with big indigenous fuel deficiencies are the countries of Western Europe and Japan. This is the situation as it exists at the moment. The basic question underlying the formulation of all long-term energy policies really boils down to our asking ourselves what we see, or expect to see, if we look forward by ten, fifteen or twenty years. Our answer to this question on the basis of present information must be that no major change seems likely to take place in the ranks of those countries that are now generally self-sufficient; that the import requirements of areas like Western Europe and Japan must, notwithstanding the development of nuclear power, be expected to grow extremely rapidly for at least the next one and probably the next two decades—so much so that the Community countries expect imported energy to account for some 60 per cent of their total energy consumption by 1980 (although this proportion will probably tend to stabilize and perhaps even dip marginally in the later years of this century); and that the big question-mark in the medium-term future is therefore what will happen to the big energy surplus and energy exporting areas of the world. Present indications are reassuring in this respect. The gap, if gap there is, will not, it must be recalled, exist or remain much beyond 1980, when nuclear power is generally expected to enter into its own on a large commercial and economic scale. Present indications are that a major industrial take-off in the under-developed countries of the world is unlikely before 1980. In these circumstances, it is overwhelmingly probable that there will be more than enough oil to meet any imaginable level of demand, certainly up to 1980, and probably up to the year 2000 and beyond, although some increases in price may be unavoidable.

The underlying and very great merit of a brief regional survey of this kind is that it at least highlights the fact that the energy supply problem is far from being a universal one. It is, on the contrary, quite remarkably localized, even if it affects the economy of both Western Europe and Japan fundamentally. The measures taken or being taken by these countries to safeguard their position will be considered later. Our more immediate problem, however, must be to look at developments in the supply situation for each of the main sources of energy, beginning with the original architect of the industrial and subsequent technological revolutions: coal.

Coal

Coal, while likely to remain an important source of primary energy in the United States, the Soviet Union and Poland, is elsewhere losing ground rapidly and seems likely to lose its place altogether in the top league of energy supplies within the next five to ten years. Thus, although its precise rate of decline in Western Europe is still undecided, the prospects for coal production being maintained at anything more than a very modest level are extremely bleak.

Coal reserves, particularly those in North America, are still immensely large. The North American continent has at its disposal

> large reserves of a wide variety of coals in sufficient quantity to support any foreseeable internal demand for any grade and to supply a large export market as well...Recoverable reserves available at present costs from thick beds less than 300 metres below the surface, total 243 million tons, assuming 50 per cent recovery. These reserves correspond to over 250 years of production at present rates.[1]

Actual coal production in 1966 amounted to 477 million tons, and present estimates are that this will increase to 529 million tons in 1970 and 771 million tons by 1980. Although coal is losing ground very rapidly in the domestic and general industry markets, a tremendous expansion is expected

[1] O.E.C.D. Energy Commission Report, *op. cit.*, p. 43.

to take place in the power station market, where coal is regarded by many public utility operators as having a very good chance of meeting competition from nuclear stations, at least during the next decade. Phillip Sporn, probably the best known American authority on the subject of nuclear competitivity, wrote in a report to the United States Atomic Energy Commission at the end of 1967 that

Nuclear power, despite a number of technological innovations, stands today substantially no further advanced in its progress toward lower costs than it was 2 years ago. Today, with increases in unit size of almost 40 per cent, atomic energy, based upon the installation of two 1,100 MW units, is competitive with coal at 22·0 cents contrasted with 24 ± 2 cents on the basis of a post-Dresden 2,800 MW nuclear unit in December 1965.[1]

Turning to the prospects of the American coal industry, Sporn added:

this very broad industry group needs badly not only to conceive and become converted to the idea of its continuing great future but also to firmly resolve to plan, build and operate—this means to produce and sell—at prices designed to assure its future. Any hope that its future can be guaranteed by Government fiat will prove as futile as governmental actions have been on behalf of coal in Great Britain and West Germany. Any one of the groups listed at the head of this item that gives up the market for conventional fuel generation undermines the efforts of the other members of the group.

Before we yield without a struggle to realize the prospect of coal and nuclear fuel becoming more constructive competitive elements in our economic system, it is necessary to recall that without question one of the great contributors to nuclear energy having reached the position it occupies today is the acceptance by the nuclear industry some 10 years ago of the fact that in coal-based generation it was facing a moving target. When the indications showed up on the horizon that nuclear energy was likely to reach that target, coal at first reacted very vigorously and very encouragingly to the nuclear competition. However, this vigour appears to have become attenuated recently. The present becalming of the sea of struggle between these two fuels certainly need not be accepted by coal, because coal has any number of directions in which to go to renew its vigour as a competitor of atomic power. It is a better competitor even now than its sponsors give it credit for. It is, to use a racetrack metaphor, a better horse than its clocking would indicate. It apparently needs a better jockey to help improve its performance.

The automation of a modern coal mine has not reached its ultimate and in the sizes that are inevitable in the coal mines of the future more automation certainly can be integrated.

The coal industry has an opportunity to make more effective use of its large

[1] The Dresden Nuclear Power Station was one of the first large scale plants to be built in the United States. With a capacity of 800 MW, it was designed to produce electricity at a cost of 24 cents per million B.T.U.s, a price that was highly competitive but not cheaper than the best coal-fired stations. It was shortly after this that the U.S. nuclear industry made its sensational bid for market supremacy with its quotation for the Brown's Ferry plant at a price equivalent to 16 cents per million B.T.U.s.

capital investment. This means around-the-clock operation of mines with three shifts versus one shift or the current occasional two-shift operation. In this effort the mineworkers themselves need to be challenged and, properly challenged, will respond.

Transportation improvement also needs to be pursued with renewed vigor and the concepts of integral trains and quick turnarounds developed to their utmost. The possibilities of transportation of coal by pipeline need to be pursued as a competitor of rail transportation. Nor can water transportation be neglected on the assumption that it already has achieved its optimum. Better barges and barges that will stay seaworthy for longer periods between maintenance, around-the-clock loading, and complete co-ordination of every step between arrival of the barge at a plant, its unloading, and its turnaround need to be optimized so as to reduce the capital cost component of this form of transportation. And electric transportation, particularly at voltages as high as 765 kV, alternating current, the highest thus far developed to the state of practice, and later higher voltages still, needs to be explored to its fullest limits.

Finally, as an important component in the struggle to maintain its competitive position, coal needs to develop an understanding of its economic Achilles' heel—price escalation. Coal needs to begin to give much thought to, and finally come up with proposals for incorporating negative escalation into its long-term contracts. The coal industry has shown no inclination to accept this concept in the few cases where it has been suggested. But it needs to recognize the possibility that over the 30-year life of the plants for which it is bidding as the energy supplier, the prospect of a reduction in nuclear fuel costs is a reality that must be confronted and met by holding out a similar prospect in the form of contractual commitments for coal resulting from improved performance.[1]

The main point that emerges from Mr Sporn's contentions—and that of many of his colleagues in the American utility field—is that the United States coal industry has, or at any rate is believed to have, a reasonably secure market for the short term and possibly even the medium term in the field of electricity generating power stations. It gives the United States coal industry, since it does not need, as yet, to ask for subsidies or other forms of governmental assistance or protection, a much greater freedom in its commercial and tactical manœuvres. The importance of this highly significant psychological factor was fully appreciated by Lord Robens, Chairman of the National Coal Board, when, during the course of his long struggle on behalf of the maintenance of the British coal mining industry, he resisted the notion of official subsidies for as long as possible. When finally he was compelled, by force of circumstances, to agree, first to a write-off of part of the industry's debts (which, in itself, was a perfectly reasonable, logical and proper step for the National Coal Board to request and the Government to grant), but then subsequently to overt Government intervention to oblige the electricity industry to take

[1] 'Nuclear Power Economics—Analyses and Comments—1967', letter from Phillip Sporn to Chairman, Joint Committee on Atomic Energy, pp. 13–14.

additional tonnages of coal—to the tune of some 6 million tons a year for three or four years—he was on a slippery slope from which the industry will have tremendous difficulty in extricating itself. It is at the time of writing difficult to see how the British coal industry will be able to forgo this assistance when the present provisions come to expire in 1971 without suffering catastrophic consequences. In the United States, on the other hand, the coal industry seems likely to continue to do sound business, if not exactly flourish, for some years to come. Some indirect evidence of this may be found in the recent spate of American coal company takeovers by the big oil and other industrial companies. Thus, in 1966, the Continental Oil Corporation acquired the second largest coal producers, the Consolidation Coal Company, with an annual level of output of nearly 50 million tons; this was followed by the takeover in January 1968 of the third biggest coal producers, Island Creek, with an annual output of some 25 million tons, by Occidental Petroleum; while in March 1968 the Kennecott Copper Corporation acquired the biggest of the American coal companies, Peabody Coal, with an annual output of some 60 million tons. Most impressive of all, perhaps, was the announcement made in April 1969 that Humble Oil Refining Company, the principal internal American subsidiary of Standard Oil of New Jersey (more widely or popularly known under its European brand name of Esso) would begin coal mining through its wholly owned subsidiary, Carter Oil Company, in 1970.

Many of the world's biggest coalfields are located in the Communist countries, particularly in China, the Soviet Union and Poland. The 1966 O.E.C.D. Report said of these countries and of some of the U.S.S.R.s smaller European neighbours that 'the measured reserves in the above countries are believed to be between 400 and 500 billion tons representing something like 200 years production at present rates'. But here also the ointment is not without its proverbial flies. Thus, the Soviet Union announced in the early 1960s ambitious plans to raise the level of its coal production from 510 million tons in 1960 to 1,200 million tons by 1980. In practice, however, coal production, after rising up to 1965, has since stagnated at about 600 million tons and it now seems increasingly probable that nothing like the original target for 1980 will, in fact, be achieved. China, as always, remains an enigma. Such output figures as we possess are highly speculative and unreliable. It is probable, however, that production will continue to rise rapidly from the present estimated level of about 300 million tons a year for the greater part of the next decade, as China continues to develop indigenous supplies to the maximum possible extent; an output figure of 1,000 million tons a year by 1980 cannot be excluded. As for the Eastern European countries, coal production has been tending to fall, with the exception of Poland, for some years and this trend seems likely to accelerate as more of the countries switch to oil and natural gas, mostly imported from the Soviet Union. Poland, the sole exception in the general decline of European coal production, still talks bravely of raising output from its 1965 level of 140 million tons to 200 million tons by 1980; even here, however, the tempo has been slackening of late and with natural gas deposits having been discovered in substantial quantities in the Eastern part of the country, the future growth of the coal industry must now be under serious review. An important consideration will certainly be the fact that, at the moment, some 25 million tons or 20 per cent of total Polish output is destined for the export market with the primary aim of obtaining foreign currency. When, however, it is borne in mind that the coal has to be hauled from the pits to the Baltic ports (a distance of some 400 km) and sold at prices aligned upon oil in some of the cheapest and most fiercely competitive markets in the world, it is manifest that this can leave the Poles with little profit at the end of the day. In all the Communist countries, there is, of course, no selling, in the capitalistic sense of the word, to be done. Fuels are produced, disposed of and consumed, according to the state of the market. From a westerner's vantage point, it sometimes seems debatable whether coal is produced because it is economic or because, for the time being at least, the countries in question have no alternative sources of energy. The recent stagnation of coal production in the U.S.S.R. and most of the countries of Eastern Europe suggests that these countries have, in their turn, become much more conscious of the economics of costs and prices and are increasingly turning their attention to other cheaper fuels. China and Poland, on the other hand, are the—no doubt temporary—exceptions that only go to underline the validity of the general rule.

The third and last of the main coal producing areas is Western Europe, where it is important to distinguish, at least for the present, between the continent and the United Kingdom. On the continent it is now almost universally recognized that the great days of coal are well and truly over. Holland plans to cease coal production in 1975; Belgium by 1980. France has recently decided upon a drastic reduction from nearly 50 million tons in 1967 to 25 million tons by 1975 and virtual extinction by the 1980s. In Germany, the position seems considerably more fluid, but the recent Government decision to bring about the creation of a single operating company for the Ruhr was arrived at with the over-riding objective for the drastic reduction in coal production from over 110 million tons in 1967 to about 70 million tons by the mid-1970s. Perhaps even more important than this general enforced reduction in coal output is the almost universal recognition by Governments, industrialists, coal producers and the trade unions that this process is not only necessary, but indeed desirable.

In the United Kingdom, on the other hand, the battle for coal, although waxing fast to the disadvantage and growing discomfiture of the coal lobby, is far from over. The Ministry of Power's White Paper on Fuel Policy of November 1967 envisaged a decline in coal production from 175 million tons in 1966 to 155 million tons in 1970 and 120 million tons in 1975. In 1980, moreover, although no figure was mentioned in the White Paper, it was an open secret that the Ministry experts were working on the assumption that coal production would not be above 80 million tons. These figures, however, were not targets; nor were they designed to be enforceable. They were simply an indication of the likely level of consumption of indigenous coal, as seen by the Government experts, in each of these years. The British coal industry, for its part, refused to accept these figures and indeed took the unprecedented step of vigorously attacking the Government's White Paper in its subsequent annual report:

> The White Paper on Fuel Policy (Cmnd. 3438), published in November, 1967 marked the transition from a two-fuel to a four-fuel economy. Within the White Paper there are really two policies—a longer-term policy for the period to 1975 based on an estimated demand for coal in that year of 120 million tons, and a transitional policy to moderate the rundown of the coal industry to 1970–1. The Board are bound to report, with regret, their disagreement with the longer-term policy. While the Board were fully consulted for the period to 1970–1, there has not been adequate discussion on the period beyond. The policy for that period would involve the displacement of such a large number of men as would be inconsistent with maintaining good industrial relations.
>
> The Board acknowledge that the formulation of a long-term policy is essential if the fuel industries are to plan the deployment of their resources of capital, technology and manpower in the way most beneficial to the national interest. Nevertheless, they would stress the importance of not taking irrevocable decisions today, in relation to natural gas and nuclear power, which will prevent the use of cheap coal in the competitive market in the 1970s. The demand estimates in the White Paper appear to be based on rates of development which drastically reduce the scope for flexibility, however much coal improves its competitive position.
>
> The coal industry, in which a large state investment has already been made, could produce a greater quantity of coal at competitive costs in the 1970s than the White Paper indicates and an unduly rapid rundown in the industry's production capacity will have extremely serious economic and social consequences. Where decisions which substantially affect the balance between competing fuels have to be taken, they should be based on a complete evaluation of the

total costs to the nation, direct and indirect, economic and social, of the various alternatives. This, in the Board's view, the White Paper has failed to do.[1]

As a further step in their campaign to maintain coal's position, the Coal Board commissioned a report by the Economist Intelligence Unit entitled 'Britain's Energy Supply';[2] this report examined the potential costs and benefits to the nation of the adoption of six different policies and in its conclusions gave a good deal of support to the coal cause. It must be noted, however, that these favourable conditions were based, as far as the economics of coal mining was concerned, on estimates provided by the Board which included a forecast that output per manshift would have risen by 1975 to between 70 and 75 hundredweights compared with 38 hundredweights in 1967. This is an exceptionally tall order, requiring an annual increase in productivity of 7·1 to 7·5 per cent, and it must be regarded as highly questionable whether anything like as high a figure as this will in fact be achieved. The closure of uneconomic pits will itself contribute little to this end since the National Coal Board's own estimates put the net benefit of these so-called negative rationalization measures at no more than 1 per cent a year. In these circumstances the Government's attitude to coal, as set out in the White Paper on Fuel Policy is unlikely to be reversed; (it is, however, possible to imagine that pressures may grow for a more optimistic official gloss to be put on the prospects for coal if only to stem the outflow of miners from the coal industry which has, on more than one occasion, shown signs of reaching alarming proportions).

The conclusion must, therefore, be that, in the European O.E.C.D. area, which in 1964 still accounted for 447 million tons of output, coal production is unlikely to be more than 200 million tons by 1980 and could well be considerably below this figure. In other words, seen against a total energy demand figure for the area of 1,370 million tons, indigenous coal would not represent more than one-seventh or some 14 per cent. In a particularly carefully worded section on the prospects for indigenous coal in Western Europe the O.E.C.D. Energy Commission stated:

> The conditions vary considerably between Member countries, but it is worthwhile to consider in general terms what the prospects for coal are likely to be in the various sectors of consumption. For example, it seems clear that the future for the direct use of coal in the transport sector is negligible. In the market which it traditionally held—the railways—electricity and diesel oil are likely to take over completely by 1980. In internal and coastal shipping, substitution of coal firing is almost complete. In industry, even the existing protection has not prevented oil from capturing many markets. It is also a sector where natural gas

[1] N.C.B. Reports and Accounts, 1967, volume 1, pp. 4–5.

[2] Published in 1968.

> is particularly attractive, and in the event of large discoveries of this fuel, it is likely to make further inroads on coal's position. The requirements of the iron and steel industry for coking coal represent a market where coal is hardly challenged as yet by other fuels. On the high rate of steel production implied by our estimates, it seems reasonable to expect that this market for coking coal, (though not necessarily indigenous coal) will continue to decline. Over industry as a whole, the use of coal is most likely to decline further. In the domestic sector too, where oil, electricity and natural gas can all offer very attractive alternatives to coal it would seem more likely that the direct use of coal should decline rather than increase.[1]

In fact, the rundown, on the evidence of the last two to three years, is likely to continue at an unabated rate and indeed has now reached such proportions that some alarm has been expressed in some quarters about the availability of coking coal (and coke) which, for some years to come, is likely to remain of vital importance to the steel industry.

Coal production in the remaining parts of the world totalled about 150 million tons in 1960; although present indications are that this figure may increase to some 250 million tons by 1980, its importance in relation to total energy demand in the countries concerned is very small.

To conclude, therefore, we may safely say that coal production is only really likely to expand in North America and in China. In North America the increase may be estimated at 250 million tons to reach an annual level of output by 1980 of 770 million tons. But while this represents a significant increase, both in absolute and in percentage terms for coal production in the United States, in global terms its effect is comparatively small, particularly when set against the declining level of output in Western Europe. Seen in relation to the total increase in energy demand that is envisaged over the period, it is of no more than marginal importance. Coal, in short, will be fighting a rearguard action for survival with its ability to do so becoming increasingly dependent upon the variations in the intensity of the competition from other fuels from one region to another; upon regional and social policy considerations, if not actual and downright imperatives; and finally, and above all, upon the readiness of governments to continue to provide financial assistance. King Coal, alas, is dying, albeit slowly, and with an old man's lust for life. As an industry it cannot be expected to make any contribution to the growth in energy supply that is confidently expected to be such a feature of the next decade; at best it may manage to remain static or even achieve a small increase in absolute terms; at worst its existence in many countries with a long tradition of coal mining behind them could well be near extinction.

[1] O.E.C.D. Energy Commission, *op. cit.*, p. 84.

Oil

The rise in the price of coal in the mid and late 1950s, 'increasing the competitive advantage of oil, caused the price of oil not to firm up but to collapse'.[1] Although this is a theory (propounded by a leading American energy economist) with which I disagree since I consider that the continuous and increasing volume of major oil discoveries in the postwar period made a collapse of the high early postwar price of oil inevitable, it provides nonetheless a useful way to introduce what has become the dominant source of energy in the modern contemporary world—dominant in terms of quantitative usage, ease and convenience of handling and multiplicity of uses.

Before embarking on an assessment of global reserves of this valuable twentieth-century commodity, it is desirable to look at what the government representatives in the Energy Commission of O.E.C.D. had to say, in 1966, about the significance of published levels of oil reserves—a point about which a great deal of play has been made by those seeking to cast uncertainty, if not outright apprehension, about the medium and long term availability of oil:

Before considering the levels of reserves, a word or two should be said about the significance of the published proved reserve figures. They do not in any sense represent a geological assessment of the limit of oil reserves. They are the result of such exploration activities as the oil companies consider it is useful to undertake. This means that they include not only the current potential which is considered to be available for expanding production; they represent the reserves which oil companies are understood to have located and which would be recoverable by known methods at present levels of costs. Finding and developing oil reserves is a costly business and, for purely economic reasons, no more would be done than is reasonably necessary. However, it must also be pointed out that in the case of the United States for example, exploitation is encouraged by considerations of national security. Outside the United States, given the present structure of the international petroleum industry, oil companies find themselves frequently compelled to apply for any new concessions available. This situation, together with the terms of the agreements relating to the concessions, has led to an intensification of exploration activities, although at the present time the extent of the currently available reserves would, in some cases, not justify the investments involved when considering the matter from a purely economic point of view, on a world wide basis. A good example of this is, for instance, the competition to acquire exploration rights in Saudi Arabia and Iran, which continues despite the considerable oversupply of crude oil. In the United States, the proved reserves have averaged some 12–15 years supply at

[1] M. A. Adelman, *The World Oil Outlook, in Natural Resources and International Development*. The Johns Hopkins Press (for Resources for the Future, Inc.), Baltimore, 1964, p. 97.

current production rates since the beginning of the century. Elsewhere, the terms and conditions under which concessions are held and operated have very often stimulated rapid development. For many years, the total world proved oil reserves have increased steadily; more oil has been located than has been produced. In considering future prospects, therefore, not only must we take proved reserves into account, but we must also allow for the possible exploitation, as techniques, costs and prices change, of deposits which are known to exist but which are not included in the proved reserves and for the existence of deposits which, from geological evidence, can be expected to be proved in the future.[1]

Broadly similar comments about the adequacy of oil reserves have been made in the long-term energy availability studies of the European Commission and in the British Government's White Paper on Fuel Policy of November 1967, although the latter did include a note of caution about the possibility of some increases in the price of oil. The oil companies have undoubtedly been subjected to increasing pressures over the last few years, by demands, often backed up by force or arbitrary measures, for bigger royalty payments by the Governments of the countries where the oilfields are located, as well as for higher taxation in some consumer countries. The oil producing countries, moreover, in 1960 established their own consortium or producer organization, the Organization of Petroleum Exporting Countries, usually referred to by its initials as O.P.E.C., with the avowed objective of extracting bigger benefits from the producing companies, and indirectly, the oil consumers in Western countries. Other factors that have tended in the same direction of narrowing oil industry profit margins, and consequently reducing their ability to finance further exploration and development programmes out of their own capital resources and reserves, have included the commercial or, in some cases, political motives behind some of the newcomers in the oil market. The Italian State-owned agency, Ente Nazionali Idrocarburi (E.N.I.) was one of the first to challenge the big majors and under its first mercurial President, Enrico Mattei, notched up some remarkable commercial successes. More recently, the French Government of General de Gaulle, faithful to a basic French policy objective of the last forty years, took measures to increase the proportion of the oil supplied to French consumers by French companies. French oil enterprises have been actively supported or where necessary newly created with strong State backing or control in order to achieve this end.[2] This has meant not only the effective rationing of market shares for foreign companies operating traditionally in the French market but an active policy of intervention overseas designed to secure French access to crude oil supplies.

Information about oil reserves in the Soviet Union and other Commu-

[1] O.E.C.D. Energy Commission, *op. cit.*, pp. 44–5. [2] See pp. 140–2.

nist countries is far from complete. Estimates prepared by the American National Petroleum Council put proved reserves in the area at the end of 1964 at 4·8 thousand million tons, of which 4·4 thousand million tons were located in the Soviet Union. The Eighth Five Year Plan provides for a production of between 345 and 355 million tons of oil by 1970, while plans for 1980 provide for a doubling of production. There is no doubt that the Russians have in the course of the last five years become very much more cost conscious and are now seeking to accelerate the rise in their oil production at the expense, where necessary, of coal. They have also shown signs of becoming increasingly interested in oil exports at about or only fractionally below ruling world prices. It is, indeed, generally recognized that the Soviet Union has little to gain, in either political or economic terms, in the short term at least, from any further depression of oil prices which must simultaneously displease Arab Governments and reduce Soviet revenue. There is, in fact, some evidence to suggest that the Communist bloc of countries as a whole may not be able to satisfy its own rapidly increasing oil requirements before the end of the next decade. If this assumption is correct, then the Soviet bloc may be faced with an outside import requirement of up to 100 million tons of oil a year. Such a development would clearly have a considerable impact upon the Soviet attitude to the desirability of supporting oil price stability. But while this is a matter that will obviously be closely watched by all the parties concerned, it is a situation with which neither the Russians nor the major Western oil companies will be confronted for a considerable time to come. In Communist Eastern Europe only Rumania is a producer of any significance, with reserves estimated at some 135 million tons. It should nevertheless be noted that this analysis of the Russian oil situation, however realistic in the short term, could conflict in the longer term with recent rumoured reports of massive Soviet oil and gas discoveries in some of the remotest parts of the Siberian tundra. According to these reports, the oil resources of Siberia would be able to support, once the necessary development work has been completed and pipelines constructed to the west, an annual production rate of between 500 and 700 million tons a year, or about as much as the whole of the Middle East put together.

As we have seen, proved reserves are normally defined as those that can be recovered at economic cost at the time that the estimate is being prepared. They are, inevitably, of a conservative nature. Thus, in the United States, discoveries of new oilfields account for no more than some 15 per cent of the annual increase in oil reserves, the balance being accounted for by upward revisions of the proved reserves of earlier discoveries.

Similar examples can be quoted from the Middle East and Canada. This position is of course likely to be dramatically changed by the recent discoveries in Alaska with total reserves now expected to amount on the basis of what are probably relatively cautious estimates, to not less than 10,000 million barrels.

The North American region, which at present provides about four-fifths of its oil requirements from indigenous resources, has proved reserves (excluding Alaska) of some six thousand million tons, equal to thirteen times the present annual production rate. Reserves in O.E.C.D. Europe and Japan are insignificant, so that both these areas have to rely upon outside sources of supply for almost the whole of their very large and rapidly increasing oil requirements.

Table 14. *Estimates of energy consumption, export potential and import requirements by regions in 1980*[1]

(In million tons of coal equivalent)

	Consumption	Exports	Imports
U.S.S.R.	2,043	50 to 150	—
China	1,845	—	—
India	201	—	50 to 100
South-East Asia	50	—	—
Australasia	119	—	—
Japan	438	—	300 to 350
Africa	157	150	50
Latin America	435	180	100
North America	2,950	50 to 100	500 to 600
Western Europe	1,492	—	625 to 780
Total	9,813	430 to 580	1,625 to 1,980
Net world import requirement excluding the Middle East			1,195 to 1,400
Bunkers			280
Total			1,475 to 1,680

If we exclude the Communist group of countries consisting of the Soviet Union, China and Eastern Europe, we are left with a situation in which the rest of the world relies for its oil supplies on three main regions: the Middle East, where the bulk of the world's reserves are concentrated, North Africa and the Caribbean area. North America is in a category of its own since it exports comparatively little of its high level of output.

[1] An Energy Policy for Western Europe, *op. cit.*, p. 2.

Proved reserves of oil are, clearly, very unevenly distributed throughout the world. The extent of the world's dependence upon the Middle East was the central theme of the 1966 publication by the Coal Producers of Western Europe,[1] in which they pleaded for greater government protection for indigenous European coal production.

Proved reserves in the Middle East, it was stated, currently amounted to between 75 and 80 years' production. While this might seem impressive at first sight it was in fact grossly misleading since it served only to make up the enormous deficiencies of most of the other producing areas when seen against the overall background of an inexhaustible world demand for oil. The true position, it was claimed, was that at the present rate of growth of demand, the Middle East's proved reserves represented no more than 22 years' production. Since the figure for the rest of the world was only twelve years, this gave a global reserve figure of no more than fifteen years.

The Coal Producers' memorandum was a document *à thèse*, designed to shore up coal's crumbling edifice within the European energy context. The final calculation as set out in the O.E.C.D. Energy Commission's 1966 Report strongly supported the contention that oil would be available in such quantities and at such prices as would enable the world's growing demand for oil, great though it was expected to be, to be fully and satisfactorily met:

On the basis of these estimates, world proved reserves are 45–50 billion tons or some 30–40 years supply at present rates of production. These estimates take no account of improvements in methods of recovery which will almost certainly take place. At present, the average rate of recovery is about 30 per cent and this could well rise to 50 per cent in the period to 1980. Such an improvement would considerably lengthen the theoretical life of reserves at present production levels without any more oil being discovered. Estimates of ultimate reserves are inevitably less precise and we have attempted none of our own. It is sufficient to note here that one estimate that has been widely discussed put the figure at ten times the level of proved reserves. For example, in North America there are thought to be about 40 million tons which should be recoverable from extensions of existing fields or in new fields thought to be discoverable under present conditions; for the U.S.S.R. an estimate—made in a recent report of the Committee on United States Natural Resources of the National Academy of Science, National Research Council (Mr King Hubbert, *Energy Resources*, Washington, 1962) gave a figure of some 27 billion tons. In addition there are the very large potential reserves in the form of shale oil and tar sands. Present estimates of oil in place from shale oil, on a world wide basis, approach 500 billion tons, most of which are located in the United States. Oil in place in the evaluated portions of Canadian tar sand deposits, which are expected to be exploited commercially at competitive prices in the near future, is estimated at 112 billion tons from which about 42 billion tons of up-graded synthetic crude oil can be

[1] See p. 102.

recovered. This latter figure alone approached present proved world crude oil reserves. The potential reserves of tar sands and shale oil represent, on conservative estimates, at least four times the proved crude oil reserves of the world. These estimates do not attempt to put a particular figure to the extent of ultimate reserves but simply serve to illustrate that the potential is very much greater than the proved reserves taken alone. In view of the progress currently being made in the conversion of coal to gasoline, coal must also be considered a potential source of liquid hydrocarbons for the future, though it would be wrong to count on any significant development in the period under review. Summing up and judging from past experience, there is evidence that sufficient further oil reserves will be developed to cope with any foreseeable increase of world demand well beyond the period under consideration.[1]

There are, however, some thoughts that require to be injected at this point into any analytical survey of the oil market, which in part reinforce, but in part also, perhaps, raise one or two questions about the description of the situation that is obtained from our look at Government thinking as reflected in the O.E.C.D.'s Energy Report. These are as follows:

(*a*) The present abundance of proved reserves has led the oil producing countries, notwithstanding the bold, sometimes aggresive, front they have adopted in the Organization of Petroleum Exporting Countries (O.P.E.C.), to seek as their primary objective the stabilization of net returns per ton produced and the establishment of new markets in preference to immoderate or indiscriminate increase in royalty payments. One of the best known academic analysts of the international oil scene, the American Professor Morris Adelman, wrote in this connection as far back as 1965:

> The underdeveloped countries are extremely concerned with price maintenance for primary commodities: it was incorporated in the United Nations' resolution noted earlier, and has been repeatedly and forcibly demanded since then, at all official and non-official levels. The former Council Chairman of the Food and Agriculture Organization has called for it as part of the world-wide struggle against hunger: 'We must de-mystify the problem...We must have done with neo-colonialism and must re-establish trade equilibrium, constantly degraded by the fall in the prices of primary materials.' The 1964 United Nations conference on trade problems will demand greater economic equality, to be achieved by price supports. The United States and Britain agree with this aim: President Kennedy has called for support of primary products prices as a basic part of this country's foreign economic policy, and the world coffee agreement fits the action to the word. Sir Eric Roll, Edward Heath's assistant in the recently lapsed Common Market negotiations, has called for more such agreements. The Council of Ministers of the Organization for Economic Co-operation and Development (O.E.C.D.) has urged 'concerted policies' to 'increase the receipts [of] the less developed nations [from] their exports, as much from basic raw products as manufactured goods.'[2]

[1] 1966 O.E.C.D. Energy Commission Report, *op. cit.*, p. 46.

[2] M. A. Adelman, *The World Oil Outlook, op. cit.*, p. 111.

(*b*) At the same time many of these countries are becoming increasingly anxious, for commercial as well as political reasons, to become associated in some way with the prospecting, exploration and development of the oil found within their borders.

(*c*) For consumer countries the most effective method of maintaining oil price stability in these circumstances undoubtedly lies in the continued and accelerated diversification of oil supplies. The main objective, however, must be not simply a multiplication of the number of countries capable of supplying oil, but rather to secure greater increases in the availability of proved reserves in areas remote from political disturbance.

(*d*) Major oil companies are faced with no alternative but to press on with ambitious exploration and prospecting schemes if only to pre-empt the field and prevent new but valuable deposits falling into the hands of newcomers with short term high profit objectives that could precipitate a temporary but ruinous crash in oil prices.

(*e*) The combination of all these factors suggests that, for the short term future at least, the relative stability of oil prices seems to have been secured. While it is impossible to give any such guarantee for the medium or longer term future, since price movements must inevitably depend, primarily, on relations between three different principals, namely the producing countries, the oil companies and the consumers, the present indications are that there will be some further increases in revenue paid to the governments of the producing countries, but that this may be largely offset by further technical improvements in transportation and distribution and storage.

Natural Gas

Natural gas is the younger of the hydrocarbon sisters. Both in absolute and in percentage terms it is much smaller and less significant than oil (or coal) and, discoveries like those at Groningen and in the North Sea notwithstanding, of far less potential impact or importance than nuclear power. It has, indeed, only emerged as one of the more important energy sources in the comparatively recent past. For many years natural gas, although intimately associated with oil discoveries, was wasted as it was flared brightly but uselessly away for lack of suitable transportation techniques. Today, with the development of gas liquefaction, giant pipelines, and vast underground storage reservoirs, the situation has changed dramatically.

Natural gas secured its first big market in North America, although the amount of gas wasted indiscriminately by flaring and by too many developers working and puncturing the same gasfields has been accurately, if sadly, described as reaching astronomical proportions. Even so, as recently as 1962, some three-quarters of world consumption of natural gas occurred in North America. Similarly, the bulk of world reserves was also located there, estimated at that time to amount to 8,500 thousand million cubic metres, equivalent to 20 years supply at the then current production rates. Undiscovered, but none the less recoverable, resources were estimated to amount to five times this figure. If natural gas reserves in South America are taken into consideration, the total natural gas reserves of the Western Hemisphere amount to some 10,950 thousand million cubic metres.

Although natural gas has figured prominently in the news in Western Europe in recent years, its actual significance in the energy market is small, amounting to no more than 2·2 per cent in 1964. Admittedly the pre-war situation when natural gas was known and used in appreciable quantities only in Rumania has been changed radically by the successive post-war discoveries at Lacq, Groningen in North Holland and the North Sea. Even so, these amounts are not comparable to those that have been discovered in the United States. Annual production from the three fields taken together is unlikely ever to exceed 120 million tons of coal equivalent in any one year. While this could, perhaps, in five or six years' time amount to a figure approaching 10 per cent of total West European energy consumption, it will never be sufficiently important to exercise a dominating influence on the market and price situation. Total reserves of natural gas in Western Europe are currently estimated at some 2,500 thousand million cubic metres.

Reserves of natural gas in Japan are insignificant and while some exploration is in progress no major new developments are anticipated. Natural gas reserves in the rest of the world outside the Communist countries and the Middle East are also comparatively small and, at some 2,700 thousand million cubic metres overall, are unlikely to have any substantial impact.

As in the case of oil, so with natural gas, the Middle East has substantial proved reserves and, it is generally assumed, vast untapped and unproved reserves. Whereas, however, technical factors previously militated against their commercial exploitation, recent developments in this respect can be expected to give it a big boost. Already a number of contracts involving the transportation of natural gas in liquefied form have been concluded between Algeria and consumers in France and in the United Kingdom, and more will undoubtedly follow. Reserves of natural gas in Communist countries were until recently reputed to be of the order of 3,000 to 3,500 thousand million cubic metres and output has been increasing at a rapid rate, reaching 220 million tons of coal equivalent in 1966 compared with 80 million tons in 1960. The largest proved and known reserves were in the Soviet Union and amounted to some 2,000 thousand million cubic metres, equivalent to about twenty years production at present rates. More recent Russian reports, however, suggest that natural gas reserves in Siberia, and inside the Arctic Circle, may in fact constitute the world's largest natural gas field, totalling as much as 10,000 thousand million cubic metres. It is, apparently, planned to increase production extremely quickly, particularly in the Soviet Union and Poland, and this is accompanied by a strong feeling of confidence that the recently discovered fields will be developed speedily, despite the undoubtedly high transportation costs from Siberia to European Russia and Eastern Europe, in order to support such ambitious expansion rates. Recent statements by Soviet leaders suggest an increase in annual production rates from some 130 thousand million cubic metres in 1965 to over 200 thousand million cubic metres in 1970 and no less than 700 thousand million by 1980. The reason for this massive planned increase appears to be, as in the case of oil, the Russians' growing cost consciousness and consequent desire to expand oil and natural gas production at the expense of their more traditional fuels.

In conclusion, it has been conservatively estimated that the world's proved natural gas reserves are of the order of 20,000 to 25,000 thousand million cubic metres. At the same time, it should be noted, as the O.E.C.D. Energy Commission pointed out, that in this particular case

> Such an aggregated figure is of lesser significance than those for the other forms of energy in view of the uncertainty of most regional estimates. What information there is, however, suggests that even those reserves that are known to exist could support a considerable expansion of the new use of natural gas throughout the world. It also seems certain that new discoveries will continue to be made, and that the new techniques of transportation have increased the number of countries in which natural gas can be competitive. But, because of the uncertainties about the size of the reserves and where they may be found, it is not possible to say with any precision how large or how fast a growth in demand they could support.[1]

[1] *Ibid.*, p. 48.

Hydro-power

Despite its local importance in Norway, the Alpine area or Siberia, hydro-electricity, like natural gas, is of comparatively minor importance in the overall energy supply and demand situation. Current production throughout the world amounts to some 800 TWh. The rate of construction of new hydro-plants appears currently to be slowing down as many of the more favourable sites have been utilized and more especially as utility companies see the balance of commercial and operational advantages swing in favour of nuclear plants. The 1968 World Power Conference estimated the maximum production capacity from all the available sites for hydro-electric installations at 5,000 TWh; the prospects of this figure being achieved within the foreseeable future are, however, negligible.

About 80 per cent of the present world production of hydro-electricity is concentrated in the O.E.C.D. area. This figure is not expected to exceed 1,100 TWh by 1980 as few economically exploitable sites remain. The main potential of further development is in Africa, South America and Siberia, traversed and dissected by great rivers whose resources have so far gone largely untapped. The Russians, in particular, are known to have ambitious plans to develop the hydro-electric resources of the Ob and Yenisei rivers in northern and central Siberia and to harness their power for the industrial development of this vast and bleak region.

Nuclear power

Nuclear power is, undoubtedly, the energy source of the future. But while this is a fact that is today almost universally recognized,[1] a great deal of uncertainty and argument continues to rage about the rate at which it can be economically introduced. In this debate the speed with which the social acceptance of nuclear power can be secured is a vital factor.

As we have seen demand for electricity is expected to rise at a rate considerably faster than that of primary energy requirements. Nuclear power is clearly the ideal form of energy to satisfy this type of demand, even if, in the longer term, it may also hold out attractive prospects for desalination and marine propulsion, particularly in large bulk carriers or highly specialized ships, such as ice breakers. Its immediate future, however,

[1] See statement by Lord Robens, Chairman of the National Coal Board, to the Select Committee on Science and Technology, Session 1966–67, p. 287: 'there will come a point in time when the decision will be only nuclear power stations for the future and perhaps a few odd stations for peak load on gas or oil. The farther that point is forward, the longer the industry has to run itself down in an orderly way and control the rundown.'

lies clearly in the field of power generation. Nuclear power's contribution to overall energy supply is today, admittedly, small. It has been calculated that in 1966 nuclear power's share of total energy availability was no more than 0·5 per cent. By 1980 however, it is expected to provide between 11 and 19 per cent of all electricity generated.[1] While by the end of this century, it will probably account for some 50 per cent of total energy availability in North America and Western Europe.

The immediate rate of demand of nuclear power depends primarily on the rapidity with which its capital and running costs can be further reduced. Comparisons of different types of generating plant, whether nuclear or conventional, should be made in terms of their costs to the generating system as a whole over their lifetime. This is difficult to do and involves highly complicated factors, such as the organization of the industry, the size of the nuclear programme and, above all, the size of the baseload and the proportion of it that is available to nuclear power.

The O.E.C.D. Energy Commision in its Report, while accepting the complexity and relative uncertainty of the situation, worked on the following assumptions:

(i) that nuclear power station capital costs would range from $150–$240 per kWh for a 400–500 single reactor station in the early 1970s and between $110 and $180 per kWh for larger stations incorporating units of, say, 1,000 MW or more by the late 1970s;

(ii) that fuel cycle costs would decrease from between 1·0 and 2·2 mills[2] per kWh in the early 1970s to between 0·8 and 1·8 mills per kWh by the end of the decade;

(iii) that operating costs including the cost of maintenance and insurance might range from 0·5 and 1·0 mills per kWh for the earlier stations to between 0·6 and 0·4 mills per kWh for the later, bigger stations.

'On the basis of these figures', the Report commented,

> assuming 80 per cent load factor (equivalent to 7,000 hours annual use at full power) and 10 per cent annual capital charges, unit generating costs might be about 5 mills in the early 1970s and 4 mills by the late 1970s. The actual values for load factor and capital charges assumed by different countries for their own assessments vary widely. The annual utilisation is highly dependent on grid characteristics and as regards capital charges they range from about 8 per cent to 13 per cent in Western Europe and from 7 per cent to 15 per cent in the United States (according to type of utility). Though some government expenditure on nuclear research, for civil or military purposes, has provided support for nuclear power development, the figures given above are based on commercial conditions.[3]

[1] W. G. Jensen, *Nuclear Power*, p. 137. Foulis, London, 1969.

[2] A mill is a tenth of a U.S. cent or a thousandth part of a U.S. dollar.

[3] O.E.C.D. Energy Report, *op. cit.*, p. 59.

The report added that, outside the United States, they saw few possibilities for coal to be able to compete with nuclear power: to do so coal producers would have to be in a position to offer coal to the power stations at prices of some $13 per ton of oil equivalent in the early 1970s, and $10 per ton by the late 1970s. The latter price is one that even the American coal producers may be hard put to achieve.

In this connection, it is illuminating to recall the costings figures given by Mr Roy Mason, the British Minister of Power, in announcing his decision in 1968 to authorize the construction of a nuclear power station at West Hartlepool, almost in the heart of the Durham coalfield. These figures, it will be recalled, showed a figure of 0·52*d.* per kWh in a nuclear power station compared with 0·70*d.* in a coal-fired power station. Since this decision was made, the National Coal Board have redoubled their claims that with coal delivered to the power stations at 3*d.* a therm or less, future coal-fired power stations making use of the fluidized combustion technique would be able to generate electricity at between 0·5 and 0·55*d.* per kWh, i.e. at virtually the same cost as nuclear stations. In these circumstances, it is argued, the use of coal could be justified on sound commercial and economic grounds and thus avoid wastage through precipitate pit closures and the millions of pounds that had been invested in the British coal industry since nationalization in 1947. The National Coal Board's claim rests, however, on two highly questionable assumptions. First, that nuclear power generating costs will remain static—or possibly, may even escalate—in the years ahead; and second, that coal costs would be reduced to about 3*d.* a therm or less. The latter price, if realized, could indeed be highly competitive with nuclear power for at least the greater part of the 1970s. It depends for its realization, however, on the stability of the domestic market for coal where consumers may pay up to almost £20 a ton for their premium smokeless fuel or anthracite requirements and in this way finance the low prices which the Coal Board hopes to be able to quote to the C.E.G.B. and other industrial consumers. The least that may be said in this respect is that it is questionable whether many domestic consumers will continue to burn solid fuels, subject as most of them are to considerable dirt, nuisance and inconvenience at such high prices a moment longer than is necessary. It is not for nothing that the Board is fighting so strongly, by means of advertisements, the development of new fuels and development of new appliances, to safeguard its market in this field, for its loss would strip it of any real price flexibility in its equally vital electricity market.

The British Government's White Paper on Fuel Policy made the following, highly pessimistic comment about the level of prices at which coal would have to be delivered to the power stations in order to be able to compete with nuclear power at various levels of development:

...estimates have also been made of the price at which coal could compete with nuclear power, that is, the price of coal at which extra investment in nuclear power would show only an 8 per cent return, which is the minimum rate of return which the Treasury recommend should normally be earned on new investment in the public sector. The 'break-even' price of coal considered in this way is estimated to come to about 2·9*d.* per therm, inclusive of transport and handling charges and...may lie between 2·6*d.* and 3·2*d.* per therm (or between 1·9*d.* and 2·5*d.* at pithead). The upper end of this range is, of course, the important figure in considering the choice between coal and nuclear power. These figures compare with...a range of 3·9*d.* to 6·6*d.* per therm with an average of 5*d.* for the present inclusive prices paid by the Central Electricity Generating Board for National Coal Board coal.[1]

Since a broadly similar reasoning may be assumed to apply with almost equal effect to the rest of Western Europe, where producers lack the large captive domestic market that is the main reserve strength of the British National Coal Board, the bleakness of the prospects for coal in the Western European area needs no underlining.

The rate at which nuclear capacity is installed during the period under review will depend mainly on the growth of electricity demand and the trend of conventional compared with nuclear costs. Electricity demand is expected to continue to grow rapidly...Even allowing for the further decline in conventional generating costs, in some areas at least, which the above estimates imply, nuclear power should be competitive for baseload in most O.E.C.D. countries by about 1970. Costs, taken by themselves, would justify large programmes of nuclear installation in the 1970s, particularly in areas lacking access to cheap fossil fuel supplies. It is unlikely, however, that the rate of nuclear installation in the early years will be determined by straight cost comparisons alone. Utilities may be expected to proceed cautiously until adequate operating experience of nuclear stations has accumulated and cost estimates have been confirmed in practice.[2]

By 1980, nuclear power capacity was expected to amount to about 13 per cent of total capacity in North America, 16 per cent in Western Europe and some 10 per cent in Japan. In terms of nuclear electricity generation, the corresponding figures were about 20 per cent for the United States, 30 per cent for Western Europe and 10 per cent in Japan. If nuclear power construction programmes are today much further advanced in the United States than in Europe or in any other parts of the world, it is doubtful whether, in percentage terms at least, the United States will long maintain

[1] Cmnd. 3438, H.M.S.O., November 1967, p. 81.

[2] O.E.C.D. Report, *op. cit.*, p. 60.

its lead. Writing of the position of atomic power in his country, Philip Sporn noted in 1967 that

Atomic power is a fact, it is established, and a somewhat new orientation of its course, I believe, is needed. It needs to be remembered that, important as nuclear energy is going to be, it still is today hardly a significant item in our economy, and even by the year 2000, with total energy use projected at around 4,250 million tons of bituminous coal equivalent, electric energy will constitute at best 50 per cent of total energy. Thus, nuclear energy (expected to account for approximately half of the total electricity produced) will account for only 25 per cent of the total energy of the country. It is, therefore, necessary that the policy that the Joint Committee of Congress pursues is consistent with a programme that will permit the other 75 per cent to live and develop to be able to do 75 per cent of the total energy job of the country. Thus the Committee, I believe, needs to see to it that the basic atomic policy fostered by various agencies of the Government does not put an unnecessary burden on the other fuels. It could even, I believe, in the interest of maintaining the competitive drive in atomic power, help coal in its effort to maintain not its position, but a position, in our expanding energy supply requirements.[1]

Sporn who, it will be recalled, is a past president of one of the largest American electricity utilities, has consistently argued in favour of a free market competition between all forms of energy, which, he believes, will act as a continual spur to greater efficiency and lower prices to the consumer. In Europe, with its expensive coal and limited supplies of oil and natural gas, effective competition in this field is limited to nuclear power and imported oil. As oil has to be imported from outside the European area and energy imports are rapidly approaching a level that many experts regard as excessive and economically dangerous, Governments are generally expected to authorize or encourage large nuclear programmes.

Elsewhere in the non-Communist part of the world, nuclear power is unlikely to make a major impact before the beginning of the 1980s, although interesting development programmes are envisaged in India and some South American countries. Their effect on the overall energy supply and demand situation will, however, be negligible. In the Soviet Union, the Russians, while they are, like the British, French, Germans and Americans, putting a great deal of time, money and effort into the successful development of a fast breeder reactor, appear to be proceeding with considerable caution in the immediate development of a commercial nuclear power station construction programme of any real substance.

The successful development of the fast breeder reactor is a factor of crucial importance for the future of nuclear power. It is not so much the lowering in costs that is at stake, since most experts believe that the existing

[1] *Nuclear Power Economics*, *op. cit.*, p. 19.

advanced gas-cooled or steam generating reactors offer exciting prospects for cost reductions, provided supplies of uranium can be maintained at about their present price level. Present proved and known reserves of uranium, obtainable at less than $10 a pound, are estimated at some 800,000 tons, sufficient to meet anticipated world demand up to about 1985. A further 800,000 tons is believed to be recoverable between $10 and $15 a pound and a further 700,000 tons at between $5 and $30 a pound. In the longer term, there are believed to be good prospects for recovering uranium from the sea (the total uranium content of the oceans has been estimated to be in the region of 4 thousand million tons) but this is not yet economically feasible on the basis of existing technology. The two immense attributes of the fast breeder are first, that it can burn plutonium, which can be much more readily obtained at competitive prices, and, secondly, and above all, that over a period of time, variously estimated at between 10 and 20 years, the fast breeder reactors will in fact produce more plutonium than they consume—a phenomenon that is usually referred to by the term 'doubling time'.

Two main comments arise from this brief survey of fuel supplies for nuclear power stations. First, as the O.E.C.D. Energy Commission stated in its Report: 'It is unrealistic to assume that no further reserves in the $5–10 price range will be discovered. Such an assumption would be contrary to previous mineral experience and it should be recalled that the rate of discovery of reserves is related to the extent of exploration, which since the 1950s has not been great.'[1] A further point to be borne in mind in this connection is the fact that even if the price of natural uranium were to double, and in the opinion of most experts an increase of anything approaching this extent is totally excluded, the effect of this on generating costs would mean an increase of, at most, 10 per cent. Secondly, the progress that has been made in the development of experimental fast breeder stations in the last two years lends strong support to the resolutely optimistic forecasts of atomic scientists, engineers and salesmen.

Other forms of energy

The expression 'other forms' of energy covers a variety of alternative forms of energy ranging from solar energy to geo-thermal and kinetic energy. To take the last of these first, the kinetic energy of the earth in the form of tidal energy alone amounts to some 13.10^{12} kWh. Various efforts have been made at various times, some of them in early historical periods, to harness the power of these natural phenomena. More recently, the French

[1] O.E.C.D. Report, *op. cit.*, p. 62.

tidal power station at Rance in Brittany, although unattractive in purely economic terms, has aroused widespread technical interest. Generally speaking, schemes of this kind have serious drawbacks in that the energy they release is excessively diffuse and requires vast and exceedingly expensive installations.

The potential or, rather, the hypothetical contributions from geothermal energy sources are, similarly, impressive. Thus, a reduction of 1 per cent in the temperature of the interior of the earth would release an amount of energy equal to the amount that would be obtained from the combustion of 2.10^{20} tons of coal; fortunately, the technical feasibility of such an operation is still beyond our ken.

Solar energy is, perhaps, the most immediately plausible of these less conventional sources of energy. Here again, the potential energy release is enormous: equal to 1 kW per square metre on a cloudless day or some 500 thousand million kW a year, recoverable either through the use of solar ovens or from the harnessing of the energy released by the physical phenomena to which the sun gives rise, i.e. wind, sea currents and tidal energy. The harnessing of the energy generated by the sun, this enormous incandescent sphere of plasma with a diameter of about 700,000 km some 150 million km distance from the earth, has aroused increasing interest in a growing number of countries.

The effective and possible use of solar energy can be divided into three main categories: its use at low, medium and high temperatures.

(*a*) Its use at low temperatures, normally at less than 100 °C, is limited to satisfying certain domestic requirements, usually in small communities. Typical of these are such uses as drying fruit and vegetables and the provision of hot water (examples of these are the 100,000 or so water heaters in Israel which, it is estimated, save the country up to 50,000 tons of fuel oil a year).

(*b*) Its use at medium temperatures, 150 to 500 °C, by means of a number of super imposed mirrors to obtain greater concentration of luminosity in solar ovens (as used in East Africa).

(*c*) Its use at high temperature, up to 3,000 °C in large solar ovens. There is a famous historical legend that Archimedes in about 250 B.C. obtained temperatures of up to 1,000 °C by means of suitably located mirrors. More modern installations such as the one at the Centre de Recherche de Montlouis in France have reached 3,000 °C. Their use lies above all in the field of fundamental research into the physical properties of refractory materials at high temperatures.

Nevertheless, when all is said and done, solar energy remains, for the present at least, a promising area of research rather than a valuable new source of energy. Its development along industrial and commercial lines in the near future is highly unlikely in the face of abundant cheap alternative forms of energy. A solar energy-powered electricity power station would require a very high initial capital investment, including the provision of stocking facilities to provide for sunless periods. These high costs would only be partially offset by the supply of free fuel and virtual absence of transport and handling costs. Finally, those countries that might for climatic reasons be more readily tempted to experiment in this direction, i.e. tropical countries, are generally lacking in the financial resources that would warrent an investment outlay of this magnitude, whose economic success would be questionable and which would, moreover, provide little additional employment.

Conclusion

Our conclusion can only be that there will be adequate supplies of energy to meet any foreseeable demand up to the end of this century. During this period, coal is destined to continue its retreat although, in view of its continuing importance in the United States, the Soviet Union and, perhaps, China, it is difficult to estimate at what stage it will effectively cease to make a really meaningful and substantial contribution to world energy supplies. It is difficult, though scarcely justifiable, not to be nostalgic about the passing of a source of energy that laid the foundations of modern scientific and technological society. But its continuation cannot be justified beyond the time needed to allow for the necessary reconversion or re-adaptation policies, either on economic or on social or humane grounds. The lot of the miner has today improved in almost all countries out of all recognition with that of his forebears of twenty or thirty years ago. It remains nonetheless a way of life that is basically incompatible with the dignity of man. With natural gas and hydro power, despite a considerable anticipated measure of growth, able to make only a modest, perhaps even no more than marginal contribution to energy demand, it will be left to oil and nuclear power to satify the world's soaring energy requirements.

There can be little doubt that for the next ten and possibly twenty years oil will continue to remain the world's dominant energy source. But during this time nuclear programmes will be developed steadily in an ever increasing number of countries, until, by the last few years of this century, nuclear plants will be providing between a half and two thirds of all electric energy generated. In a world that is becoming increasingly dependent upon electric energy, the impact and importance of nuclear power cannot but grow. But this is for the future. The immediate task confronting governments in many parts of the world is essentially one of transition and of finding a solution to the manifold economic, social and regional problems of replacing the old and faithful obsolescent energy industries of the past, like coal, lignite and blast furnace gas, with the 'white-heat' fuels, i.e. oil and, even more especially, nuclear and, in the distance, thermo-nuclear power, of the twentieth century and beyond.

SECTION V

THE CASE FOR NATIONAL AND INTERNATIONAL ENERGY POLICIES

The balance between supply and demand

The conclusion which emerges, inexorably, from our consideration in Section IV of energy supply and demand is that the world is moving towards, and has indeed already partially reached, a situation of plentiful availability. The sum of all the various forms of energy is such that the development or appearance of a fuel shortage is now totally inconceivable. The question that arises today is no longer one of how the massive volume of energy demand will be met (as was certainly the case in the middle 1950s), but what parts should be assigned—whether it be arbitrarily or through the undisturbed play of market forces—to the several competing sources of energy. It is of course a question to which one can envisage many answers, according to the differing circumstances of all the countries concerned. As the O.E.C.D. 1966 Report put it: 'In practice, for each Member country, energy policy involves achieving a balance between different and sometimes conflicting objectives; and each country seeks to establish this balance in the light of its own resources and interests.'[1] Nevertheless, there are two objectives which, it can be safely said, apply to all countries. These are, first, the assurance of adequate and reasonably secure supplies of the various forms of energy that are needed to sustain economic growth; and, secondly, the assurance of reasonable energy prices and the encouragement of reductions in the cost of energy both to the community as a whole and to individual private consumers. At the same time it is impossible to avoid the conclusion that, in the vast majority of countries, governments have usually in the past tended to base their energy policies on short rather than long term considerations. At the same time, they have shown a marked unwillingness to take the economically somewhat venturesome but socially unpopular measures which it is often impossible to disassociate from a coherent, logically constructed and forward-looking energy policy.

[1] O.E.C.D. Report, *op. cit.*, p. 103.

If we consider first the measures that have been taken during the last ten years or so by the governments of leading industrial countries with a view to guaranteeing the availability and security of energy supplies, we are confronted by a formidable phalanx of rules, provisions and arrangements, ranging from the encouragement of exploration and development of energy resources (by such means as the granting of tax privileges) to the protection accorded to various forms of indigenous production, the diversification of sources of supply and the creation of stockpiles.

Taxation is undoubtedly the choicest instrument, with an almost infinite potential, yet developed by the art of government. Fiscal measures of one kind or another are widely used to encourage indigenous production, to penalize imports or to reinforce the competitive position of one or more types of fuel: typical are the depletion allowances of the United States to stimulate exploration for oil and natural gas. In one of its forms 'percentage depletion' allows a producer to deduct up to a stated percentage of gross income from oil and gas production from federal taxable income. Similar but lower allowances apply to coal and lignite, oil shales and fissile materials. Another important tax provision allows drilling and development costs in oil and gas exploration to be charged against current income. Canada, France, Germany, Spain and Turkey are among other countries that also give substantial fiscal incentives to promote indigenous energy production. Indeed, the French system of depletion allowances goes one further than that of the United States and stipulates that exempted profits should be reinvested in further exploration.

By far the most significant means of support up to the present time, however, has been the mantle of protection thrown over various forms of indigenous production—though this has meant, in the main, assistance for indigenous coal in Europe and Japan and indigenous oil in the United States—to protect them from the otherwise all-conquering blasts of international energy trade competition. At the same time vast sums have been spent, usually through military budgets, in research and development work on nuclear power, and the construction of nuclear power stations required to produce the supplies of plutonium needed for military purposes. The arguments in favour of protection of indigenous sources of supply are many and varied, but, like all arguments for any policy, are also open to doubt and criticism. It is, for instance, a geological as well as an economic fact that the conditions in European coal mines are such that the coal produced in them cannot be sold at prices that are at once economic and able to bring in a reasonable return on the capital invested in them unless they are very heavily subsidized by the national Exchequer.

To give a few examples, we have set out below the current effective levels of government subsidies, direct and indirect, in the four coal producing countries of the Common Market:

I. *Belgium*	£ million
Covering of social security expenditure	47
Covering of operational losses	27
Depreciation and capital assistance	3½
Miners' annual bonus	2
Miners' holiday and housing benefits	2
Coking coal subsidies	2
	83½

II. *France*	
Covering of social security expenditure	107
Rationalization aid	50
Reduction of interest charges on borrowings	10
Contribution towards coal research	1
Coking coal subsidies	7
	175

III. *Germany*	
Covering of social security expenditure	288
Closure premiums and tax reliefs	37
Servicing of decentralized stocks	2½
Tax reliefs to German power stations to encourage the purchase of German coal	10
Price subsidies to increase use of German coal in power stations	5½
Subsidized rates for coal transport	6
Grants towards building of coal-fired group heating installations	1½
Share of cost of miners' wage increase	12½
Miners' shift premium	12½
Compensation for unworked paid shifts and losses due to short-time working	20
Coking coal subsidies	16
	411½

IV. *Holland*

Covering of social security expenditure	7½
Interest on re-organization grants earmarked for future payment to private mines	1
	8½

Or, for the four Common Market countries as a whole in 1967, a grand total of £678½ million. In view of the fact that total coal production in the Community in 1967 amounted to 189·5 million tons this was equivalent to an overall subsidy of 71/7 a ton, ranging from 20/8 a ton in the Netherlands to 70/6 in Germany, 73/6 in France and 101/10 in Belgium.

These clearly are staggering and enormous figures. But they do not constitute the only forms of protection for indigenous energy production, which also include import restrictions and duties, taxation of selected forms of energy and indirect protection.

If we look for a moment more closely at just one of these—the import restrictions and duties that are enforced by a large number of industrialized countries—we can see at once how formidable a barrier has been erected in the course of the last ten to twenty years against the threat of free trade in energy. In the United States, the measures taken to safeguard the maintenance of indigenous oil production started in the middle of 1957 with a 'voluntary' restriction of crude oil imports. An official report prepared in 1957 had stated that

> ...if the imports of crude and residual oils should exceed significantly the respective proportions that these imports of oils have to the production of domestic crude oil in 1954, the domestic fuels situation could be so impaired as to endanger the orderly industrial growth which assures the military and civilian supplies and reserves that are necessary to the national defense. There would be an inadequate incentive for exploration and the discovery of new sources of supply.

Quotas at that time applied only to the area to the east of the Rocky Mountains, but were extended six months later to the west coast. Since, however, these initial import controls applied to crude oil only, there soon emerged a marked increase in imports of refined products. The result was that when import controls were made mandatory in 1959 they were applied to oil products as well as crude oil imports. Set initially at 9 per cent of estimated demand, the import quota was raised in 1963 to 12·2 per cent of anticipated indigenous production of hydrocarbon fuels. This system operated smoothly only until 1965. In that year the United States Department of the Interior authorized a special quota (estimated to be worth approximately one dollar per barrel of crude oil) for the Phillips Petroleum

Company to import crude oil and semi-refined products for a large petro-chemical plant which it was building in Puerto Rico. This, inevitably, opened the door to a whole cascade of requests for similar special treatment by other companies. As a result, while the present situation is that the 12·2 per cent limit continues to remain in force, there is now a whole gamut of opinions and pressures ranging from the extreme view, cherished by the domestic American oil producers, that the present quota system should be maintained and the special quotas phased out as their terms expire, to the independent petrochemical manufacturers who want complete freedom to import. Commenting on the U.S. system of import controls and so-called 'prorationing', Professor Adelman said as far back as 1964 that 'prorationing is generally recognized as price-fixing, but even more basically it is featherbedding. Regulation has sacrificed the more efficient to the less efficient output, and has sacrificed the industry as a whole to the local supplying population to promote jobs and incomes within the state.'[1] In 1969 Mr Cordell Moore, Secretary of the Interior in the Johnson Administration, giving evidence to the U.S. Senate Anti-Trust Committee, estimated the cost to the American consumer of import controls at between $1 and $2 thousand million a year.[2] He agreed that the

[1] M. A. Adelman, *The World Oil Outlook in National Resources and International Development*, published for Resources for the Future, Inc. by Johns Hopkins Press, 1964, pp. 57–8.

[2] An article published in Resources for the Future in 1968 stated that: 'The close supervision over the total volume of imports is affected by allotting specific quotas to domestic refiners on the basis of their refinery runs after allowance for certain historical quotas. The quotas are staggered in such a way as to allot higher proportionate quotas to smaller refiners, in line with the "small business" policy of the government. Interior refiners do not in fact process their quotas, since it is uneconomical to do so, but trade them to East Coast refiners in exchange for domestic oil—a transaction that is estimated to yield an average profit of about $1·25 on each barrel of quota traded. The essence of the quota system is, to a high degree, that the large international companies import and process oil they have produced in other countries, but pay for the privilege by buying the quotas of refiners who process only domestic oil. The system is a profit-sharing scheme which forces East Coast refiners to share the benefits of low-cost imports with inland refiners.'

Such a policy is expensive. An official report to the President by the Petroleum Study Committee in 1962 estimated it then at $1 a barrel or more than $3½ thousand million at the then rates of consumption. A high price to pay for security, and one that makes the cost of the European coal support measures look like very small beer. The suggestion has been made that one possible solution would be to provide the U.S. armed forces with special federal reserve areas, like those earmarked for the American navy, or, alternatively, to charge part of the additional cost to the military budget. But while such a solution would have the merit of making oil cheaper to the American consumer, the cost to the country (since the concessions would presumably have to be purchased at current market prices from the oil companies that presently own them) would of course remain.

cost of oil would, if imports were freed, drop to some $2 a barrel in the short term, but also argued that this would result in a loss of indigenous production of over 500 million barrels a day. If the wells affected were to be kept closed for any extended period of time, this would mean a total loss of some 6 thousand million barrels or some 15 per cent of total U.S. oil reserves. Other consequences would be the closure of small inland refineries and depression for the coal industry. In the longer term, moreover, there could be no guarantee that oil prices might not in fact rise to above the current level.

Fundamentally, however, while some government relaxation with regard to oil imports seems inevitable, the amount of oil imports that will be tolerated will remain, in the last resort, a factor of official American strategic thinking, based in large part on the maintenance of indigenous oil production at a level capable of sustaining the country's requirements in a time of emergency. The end result for the big American oil companies, however, is the maintenance of a privileged market, where foreign competition is very much a minority factor and where the sales of their relatively expensive home produced oil is, virtually, guaranteed by government action. It should, however, be noted that one of the first actions of the Nixon administration in February 1969 was a decision to review the validity of United States oil import policy and to transfer responsibility for its formulation from the Department of the Interior to the President's own office.

Prorationing in the United States goes back in fact some forty years. By the late 1920s a situation had developed in which countless oil producers, many of them working the same oilfield, and by sinking far too many wells recklessly wasting large quantities of oil and natural gas, were producing far more oil than the market could take. This, inevitably, had catastrophic consequences on price levels and profitability margins. As a result first one state, Oklahoma in 1928, followed almost at once by Texas and then by all the other major American oil producing states, introduced a system of production quotas for all producers in each state. The control exercised in the prorationing system was strongly reinforced in 1930–1 when great new oilfields were being developed in Oklahoma City and East Texas which vastly increased the amount of oil that was made available to the market.

The last word on the effectiveness or otherwise of the present U.S. system of oil import controls can best be left to Professor Adelman:

> The present incentives situation is, so to speak, the algebraic sum of different elements. Two federal policies, tax benefits and import restrictions, are directed

towards maintenance of relative self-sufficiency. While they undoubtedly contribute towards this end, there is no evidence concerning the degree to which they do so. The force of the incentives provided by the federal policies cannot be judged independently of the incentive structure contained within the state regulatory systems. These systems are not by design directed to the stimulation of exploratory activity. They are directed to market stabilization, adjustment of property rights, sharing of production among all producers, and waste avoidance in a limited sense.[1]

The tasks confronting the Nixon task force on oil policy were formidable and the areas of inquiry defined as follows:

1. Impact of oil import controls on the 'incentive to expand domestic reserves'.

2. The cost of controls to the consuming public.

3. The impact of the removal of controls on the profitability of wells currently in production.

4. The effects of the present control system on the operation and future development of the petrochemical industry in the United States.

5. The implications of various alternative systems of control on national security, international relations and balance of payments considerations.

6. The implications of the new and possibly immense oil discoveries on the north shelf of Alaska and in other parts of the North American continent.

7. The administrative aspects of controls and the way in which any necessary controls can or should be most efficiently and equitably operated.

Even American energy policy, it is recognized however, cannot be decided upon in isolation. Whatever decisions President Nixon may finally take, they will have a profound effect upon the United States' main energy suppliers, Canada and Venezuela. Some indication of the gravity of the situation for Canada was given by Premier Trudeau's hasty flight to Washington in March 1969 to discuss North American continental oil and energy policy with the American President. In a Press conference after that meeting Premier Trudeau said notably:

> We have a continental oil policy of sorts. It was set up in the past, and it worked reasonably well. The technical details of it are perhaps a bit elaborate, but essentially it means that Canadian oil producers sell to western Canada and sell (also) to the United States an amount roughly equivalent to that amount of oil that eastern Canada purchases overseas and, notably, from the Venezuelan producers. It is a deal between the American government and the Canadian government which is cost-saving for both parties.

[1] See U.S. Energy Policies. An agenda for research. Published by Resources for the Future, 1968, p. 49.

The new oil discoveries and the implementation of this past policy is creating problems. We did discuss them and we are announcing ...that there will be further meetings...with a view to looking at this continental oil policy and discovering the new avenues that might want to be followed.

I think we have arguments for the United States in the sense that our oil is not only cheaper, but it is more secure in terms of defense in any future conflict. It is continental oil. It is more easy of access. And if we do not continue exploring and discovering new sources of oil, there might come a time when there will be an oil gap that we won't be able to fill on this continent.

Discoveries at Prudhoe Bay (Alaska) perhaps retarded for some years the development of such a gap, but I think it is very present in our mind, both the American government and we will now be seeking to establish new guidelines for a policy which will be in the mutual interests of both countries to permit the encouragement and development of oil resources in Canada, and at the same time not disrupting your internal markets.

We find that the discussion went very well, that there was a great deal of understanding between our governments on the overall aims, and we are very optimistic that there will be emerging a renewed oil policy which will be satisfactory to both governments.

The Canadian worry, in a nutshell, was that the big discoveries in Alaska should not be allowed to interfere with Canada's longer term oil trade prospects with the United States. For the Canadians a firm American commitment, preferably within the framework of an overall North American continental oil policy, is a matter of crucial importance. For the Americans, on the other hand, the long term proximity of big, secure and fairly cheap Canadian oil supplies is an important strategical factor which no American administration, whatever its economic thinking, is likely to overlook.

In Western Europe and in Japan it is coal that is the main beneficiary from protectionist policies. Thus, in addition to the direct financial assistance payable to the coal industries of these countries, there is in operation a wide range of other measures designed to reinforce coal's crumbling position. These measures include restrictions of outside and cheaper coal imports, either by means of quotas or licensing procedures designed to achieve a similar effect: heavy taxation of oil, i.e. DM. 20 a ton in Germany; special facilities for coal transport; and, above all, government intervention, particularly with nationalized industries that are big energy consumers, with a view to securing their agreement to consume substantial minimum quantities of indigenous coal in return for financial, fiscal or other concessions. In Japan, there are special arrangements laid down by the government designed to ensure that the high cost indigenous coal production is fully utilized.

Protection is, of course, difficult to reconcile with the achievement of such desirable objectives as cheap energy and free consumer choice. The O.E.C.D. 1966 Energy Report put forward the following explanation:

> The reasons for giving protection are complex. They include the demands of adequacy and security of supply, both from the point of view of quantities and prices, reasons of regional economy, and those of the balance of payments. There is also the fact that coal mining is by its nature inflexible; particularly in view of the large labour force and considerable amount of capital employed in it, governments are concerned that its rationalisation should be orderly and should not involve serious social and economic consequences. By raising the price of energy and creating a less efficient use of resources, there are disadvantages in artificially maintaining non-competitive sources of production and the measures of protection applied are usually the result of difficult options and delicate compromise.[1]

The second main line of policy pursued by governments to safeguard their energy supplies has been to diversify the sources of supply to the greatest possible extent. Despite the emergence of nuclear power this means in practice the diversification of oil supplies. This has taken a number of forms. First, there has been a great increase in the number of commercial oilfields in operation throughout the world. Whereas, for example, Europe tended in the early and middle 1950s to rely almost exclusively on the Middle East for its oil supplies, today substantial amounts can be, and are, drawn with comparative ease from Libya, Algeria, and Nigeria, while the number of oilfields in exploitation in the Middle East area itself has increased severalfold. In the longer term oil may well flow to Europe and to the north-east coast of the United States from Alaska, particularly if the experiments now being carried out to transport oil by tankers through the North-West Passage round the northern coast of Canada are successful. Second, there has been a substantial increase in the number of countries and, consequently, of oil companies engaged in the fields of prospecting, exploration and production. This development, however, has been accompanied by important political as well as commercial repercussions. Thus, the decision of the Japanese Government to promote the creation of two national oil companies by means of low interest loans, while it has not had an important bearing on the physical work of oil exploration and production in the Far East and Arabia, at least up to the present time, has had international repercussions as a result of these companies' approach to the question of royalty payments to the countries where the oilfields are located. The same considerations apply for the operations of the Italian state-owned oil agency E.N.I. (Ente Nazionale Idrocarburi) and, more recently, to the

[1] O.E.C.D. 1966 Report by Energy Commission, *op. cit.*, p. 109.

French E.R.A.P. company in its deal with the Iraqi Government. Previously, a signature of contract between owner governments and the oil companies signified, in effect, that the latter paid for the oil they discovered by an exploration premium which was actually payable regardless of whether oil was discovered or not, by the acceptance of substantial exploration risks, and by accepting an obligation to take up and sell any oil that was discovered. Further typical clauses normally incorporated in such arrangements provided for:

(*a*) a 50 per cent participation by national companies of the countries concerned;

(*b*) the acceptance by the oil company concerned of all research costs;

(*c*) the delivery of the crude oil due to the national company of the host country at a 'half-way' price (i.e. at cost price + taxes + half of the profit);

(*d*) the payment of a premium upon signature of the contract.

The deal which E.R.A.P. signed in 1968 with the Iraq National Oil Company was a completely new type of agreement, in which they undertook to act as entrepreneur for the Iraqi national company. Under these arrangements the Iraqis remained the sole owners of any oil produced, although it was left to E.R.A.P. to select the most promising oil-bearing areas. If and when oil was actually discovered, it was to be sold by E.R.A.P. on behalf of the Iraqi National Company at the ruling market price. E.R.A.P.'s return comes in the form of a contract for a fraction of the total quantity of oil sold at a price marginally above cost price (and, presumably, therefore, well below the market price). The E.R.A.P. deal was in fact doubly significant because the formation of the Iraq National Oil Company was the result of the Iraqi Government's arbitrary seizure of control in 1964 of all the concessions previously granted to the international grouping operating under the name of the Iraq Petroleum Company which were not at that time being exploited. These, it was claimed, amounted at the time to over 90 per cent of the company's concessions and included the very rich oilfield of Rumalia in northern Iraq.[1] The French also made a bid at about the same time through another state-controlled oil company, the Compagnie Française des Pétroles (C.F.P.), a participant in the Iraq Petroleum Company, to obtain concessions in the Rumalian field, but without success; the concessions being granted by the Iraqi government to the Russians.

[1] The part of the Iraq concession expropriated under the Iraq Government's Law 90 was 90 per cent of the total, and included North Rumalia, in the northern end of the Rumalia field.

Deals of this kind had obvious political undertones, as had the five-day official visit to France in April 1969, shortly before President de Gaulle's resignation, of the Kuwaiti Prime Minister and Crown Prince, and the signing of an agreement for the establishment of a French–Arab investment bank. Kuwaiti spokesmen stressed the fact that their country had long been a firm believer in the principle of strict reciprocity in foreign relations according to the attitude of the country concerned towards the Arab–Israeli conflict. The Franco–Kuwaiti agreement was in line with this principle. Replying to questions by a correspondent from *Le Monde*, the Crown Prince is reported to have stated that

> I have had an opportunity to exchange views on this subject [i.e. French concessions in Kuwait] with M. Pierre Guillaumat, the President of E.R.A.P. I explained to him that our territory covers only a small area and that we are not at present in a position to offer concessions. Things will, of course, be different when one of the oil companies already established in our country relinquishes part of its exploration area. We have, however, agreed that E.R.A.P. could enter into association with Kuwait in undertaking operations for the extraction of raw materials in other Arab countries. An agreement in principle has already been reached on this matter.

The commercial aspects of the form of remuneration agreed between E.R.A.P. and the Iraqi national company would probably not appeal to many oil companies, even though the initial capital stake required is much smaller than in the more conventional or classical types of contract in the international oil business. The activities of the French oil groups in this respect are, however, governed very largely by political considerations and objectives and it is to them that one must look for an explanation and justification.

An even more commercially venturesome deal, however, was to follow in March 1969 when the American Continental Oil Company signed an agreement with the Iranian National Oil company, under which Iran will reportedly receive about 90 per cent of any profits arising from the venture. Under this deal Continental Oil must invest at least $8 million in exploration and other operations over the next five years and then return 50 per cent of the concession area to the Iranians as a contribution to the country's national reserves. Subsequently, Continental Oil must invest another $2 million in the following two years before returning a further 25 per cent of the original concession to the Iranians. At the end of the seventh year Continental Oil has an option to scrap the agreement on condition that, before doing so, it invests another $4 million in exploration costs and pays $3 million in bonuses. Nor is that the end of the story. There are still other provisions for payments, in cash or in oil, to the Iranian government—sufficient to indicate that the governments of the countries with oil in their subsoil are determined to exact an exceedingly handsome return on their deposits. The currently fashionable word, indeed, is 'participation'. By this is meant the involvement of the host countries not only in the exploration and production of the oil in their subsoil, but also in its overseas sales, though how this is to be done remains a commercial and economic mystery. It represents nonetheless a very real and deeply felt desire on the part of many countries to have a greater degree of control over the ultimate disposal of the oil on which their economy is often almost entirely dependent. Participation is an objective that is therefore political as much as, if not more than, economic or commercial. Most of the countries concerned, on the other hand, are also very conscious of the fact that there is a point beyond which they cannot drive the oil companies without sparking off a confrontation of such magnitude that it could well, ultimately, completely shatter the existing order of relationships between the host countries and the big international companies. Few would want to see matters reach this stage. Another alternative which apparently has strong advocates within the National Iranian Oil Company, is for national companies of oil-producing countries to look into the economic and other related arguments for moving into the downstream business.

In the third place, there are the measures being taken by a number of governments, such as those of France, Germany and Japan, designed to ensure that at least a definite minimum part of their home market is kept for national companies. The French Government has once again been particularly active in this respect. This tendency towards national self- or at least part-sufficiency in the oil industry has shown some signs recently of becoming contagious. Thus, the German Government, towards the end of 1968, acted with great firmness in resisting the bid of the French company, Compagnie Française des Pétroles (C.F.P.), to acquire control of the German Dresdner Bank's holding in the German Gelsenberg Oil Company. The background to this incident was the German Government's concern at the extent to which the German oil market was falling into the hands of foreign companies. It was estimated that at the end of 1968 only some 25 per cent of the market was still under German control and this was divided among a number of companies, some of which seemed unlikely to resist the lure of selling out at attractive prices to foreign companies. It was at this stage that C.F.P. made a strong bid for the Dresdner Bank's minority holding in Gelsenberg, the only German oil refining company with worthwhile overseas concessions. The deal was on the point of conclusion when the German Government intervened and insisted that Dresdner's holding should not be sold out to a foreign interest. After some intense economic and diplomatic activity behind the scenes it was eventually announced that Dresdner's interest would be taken over by the Rhein-Westfalische Elektrizitätswerke, Germany's largest producer of electricity. The C.F.P. bid, despite strong support from the French government, had been foiled; but, as it was pointed out, the French had little grounds for complaint as the Germans were only seeking to do, on a smaller and more modest scale, what the French had set out to do themselves within their own frontiers. A further indication of the German government's desire to strengthen its indigenous oil industry was the support it gave, at about the same time, for the merger between the country's fourth largest crude oil producer, Gewerkschaft-Wintershall A.G. with the chemical giant Badische Anilin und Soda-Fabrik A.G. It was not long after this that the Federal Government announced its decision, in February 1969, to contribute DM. 575 million towards the formation of a German national oil organization. Talks between the Federal Government and the country's eight largest oil enterprises were to start at once. The national oil organization (later baptized Deminex) would not only trade under its own name but would also, as a subsidiary of the eight founding companies, prospect and drill for oil and negotiate oil agreements with foreign countries.

Another point worth recalling in this connection is the limitation operated by many countries on imports of oil from Soviet bloc countries. These are almost invariably negotiated through bilateral trade agreements and are inevitably subject to close government supervision and control. Thus, the Common Market countries, for instance, held informal discussions some years ago about limiting their oil imports from the Soviet bloc countries to 10 per cent of their total requirements, while the United Kingdom does not at present permit any imports from this source. Although most of the oil deals with Soviet countries are part and parcel of much broader trade barter or exchange agreements, they have in the process the dual effect of minimizing dependence upon this particular area of supply and of further increasing the overall regional diversification of supplies.

A last means of promoting the security of oil supplies consists of compulsory stockpiling by means of legislation. The Common Market countries—following an O.E.C.D. lead—agreed in 1968 to maintain a level of oil stocks equal to 65 days' supply. The most notable example, in this field, however, is provided by the United States where the so-called 'naval reserves' (which are, in effect, as we have seen, federally owned lands under the control of the United States navy) contain at least 165 million tons of proved reserves which can be made available at short notice for military and industrial emergency requirements. As the O.E.C.D. drily commented: 'the North American conservation regulations not only serve to avoid physical waste but also result in a considerable reserve capacity of oil and gas production'.

Before proceeding any further in our study of government measures in the field of energy policy, it may be timely to recapitulate briefly on what we have established so far. It is, of course, important to remember at all times that the task of all governments is to achieve the most favourable balance between its often conflicting objectives and requirements.

> In working towards this objective government action can be directed both at lowering energy costs and at influencing prices. Action taken to decrease energy prices or costs on a particular market or for certain consumer groups need not have the same effect on costs for the community as a whole. The effect of some measures is only to shift costs or fiscal charges between different consumers or between consumers and the budget, while other measures may actually contribute to lowering the basic costs of energy. Special tax privileges and price fixing for the benefit of certain consumers fall into the first category; encouragement of rationalisation to reduce costs and government-sponsored research belong to the second group.[1]

We must now look at each of these categories in turn in greater detail.

As far as price control or price fixing is concerned, examples of complete control have been rare outside the Soviet bloc. Government control has been applied on several occasions in the United Kingdom when the government imposed price restraint on the National Coal Board, particularly during the first post-war decade when coal was in short supply and could undoubtedly have commanded a considerably higher price than was in fact the case. Similar examples may be found in France where the nationalized Charbonnages de France, like the Coal Board in the United Kingdom, must seek government approval for proposed price increases before putting them into effect. In a slightly different sphere, namely that of natural gas, the Dutch, French and British governments have all played an important part, directly and indirectly, in fixing the price at which natural gas should be sold within their own borders. The Dutch government, for instance, not only determined the cut-price rate at which natural gas from Groningen should be made available to the new industries which it wished to develop in the industrially underdeveloped region of northern Holland, but also kept a watching brief over the commercial policy of the official natural gas sales agency, Gasunie, in which it

[1] *Ibid.*, p. 113.

was directly represented, to avoid any excessive disturbance in the overall Dutch energy market. In the United Kingdom, it is hard to believe that the Gas Council would have secured the extremely favourable price of 2·87 pence per therm for the natural gas it buys from the oil companies operating in the North Sea but for its monopoly position as a buyer and, at the very least, indirect encouragement from the government to stand firm. In the United States, the gas and electricity industries are partly under the control of the Federal Power Commission which exerts its influence through the regulation of interstate operations, and partly under the control of the State governments. There is also the fact that the Interstate Commerce Commission controls charges for the transport of other forms of energy.

There are also in many countries provisions for governments to take direct action in the event of an exceptional situation developing. The O.E.C.D. Report stated in this connection that many governments

> have tried to ensure that the adjustment of supply and demand should also take into account foreseeable long-term trends, in order to avoid irrevocable decisions being made in response to the short-term consequences of cyclical fluctuations in the economy or other particular circumstances. For example, Japan's Petroleum Industry Act, promulgated in 1962, authorizes the Ministry of International Trade and Industry to impose standard selling prices for petroleum products if serious price changes threaten to jeopardize the objectives of securing a regular supply and of keeping prices low and reasonably stable.
>
> Summing up, direct government control of energy price has recently been less prominent than in the past. It has probably not influenced the general level of energy prices to a large extent. However...governments have considerable influence on prices through measures less direct than full or partial price control.[1]

More important from an overall economic point of view is the inflationary effect that the operation in a number of countries of official government policies, designed to protect indigenous sources of energy production, has had on energy prices in general. The O.E.C.D. Report noted in this respect that 'practically all countries with growing awareness of the contribution of energy to economic development, have attempted to reduce the cost of energy to the community as a whole and to the consumer by stimulating technical progress and rationalisation and by encouraging growing efficiency of energy usage.'[2] But there are, obviously and inevitably, limits to what can be achieved in this somewhat undynamic field of research.

There are, of course, enormous differences in attitudes and approach

[1] *Ibid.*, p. 115. [2] *Ibid.*, p. 113.

between countries with an important element of expensive indigenous energy production and those without. The Common Market countries, in their energy policy discussions, have provided a fascinating example of this dichotomy in their objectives. Whereas, for example, in their common agricultural policy, they are pursuing an ultra-protectionist line of policy in order to provide France with her pound of European flesh as a payment on account for her consent to the overall industrial policy of the Community, the Common Market's energy policy thinking has been marked, on the whole, by fairly strong liberal intentions and inclinations. The Germans, with their important coal interests, are in something of a cleft stick, but will probably in the long run accept to pay the costs required of them in order to achieve a cheap energy policy as a sacrifice on the altar of the overriding interests of the general and basic industrial manufacturing sectors on which their strength and prosperity depend. It will be interesting to see whether their growing industrial might will, again in the long term, give them the necessary confidence to call—as reason would seem to indicate—for a liberal agricultural policy and so make the countries of the Common Market accept the role of the dynamic power house in stimulating world trade that history, geography and destiny, among them, would appear to wish to thrust upon them.

But it is time to return, as the French say, to our muttons: in this instance, the more specific measures affecting particular consumer groups. We have already seen the high if illogical volume of subsidies for coking coal paid by the Community governments to their steel industries—illogical since coking coal supplies from overseas producers at the low prices represented by these subsidies are, in fact, largely unobtainable. Other examples are provided by the tax reliefs granted in some O.E.C.D. countries for energy supplies to the electro-metallurgical and petrochemical industries. In the transport sector governments, while usually applying high taxes on private motorists to obtain fiscal revenues, permit lower taxation for diesel fuels for industry and partial or complete remission from tax for agriculture and fisheries.

It must be recognized nonetheless that, on the whole, even the more protectionist countries have continued, as far as their objectives would permit, to encourage competition between the various sources of energy:

All oil consuming countries have a strong interest in maintaining a reasonable level of oil prices on a long-term basis and preventing wide fluctuations in prices. Both the private oil companies and the exporting countries also have an interest in achieving these objectives, although of course the latter are seeking higher returns, and conflicts in this, as in every other field, are sometimes inevitable. As pointed out in the last report by the O.E.C.D. Special Committee for Oil, Member countries are 'vitally concerned with the maintenance of good relations between oil producing countries and the oil companies'. They are further concerned with the maintenance of good relations between themselves and both exporting countries and oil companies. But this is a subject which goes well beyond the field of energy.

A competitive market structure serves the interests of both producing and consuming countries, for its existence tends to ensure that supplies are made available at prices which reflect the economic conditions of supply and demand. It further helps to ensure that neither producing countries nor oil companies nor for that matter importing countries, acquire a dominant position that could be abused. The principles in this and the preceding paragraphs are generally accepted, although it must be recognized that, as in most cases in real life, the actual situation within energy markets does not fully correspond to the theoretical ideal.

The United States actions in this field illustrate the efforts of governments to develop a competitive structure. It has encouraged the development not only of indigenous oil and natural gas resources, but also the development of access by American companies and associates for foreign resources. At the same time the authorities have encouraged the development and progress of independent companies in both domestic and overseas oil activities as a means of balancing what might otherwise be a controlling position of a few major groups. The first application of the 'Sherman Antitrust Act', was in fact directed at the oil market, which in 1911 seemed liable to come under the influence of a single company. The national oil companies in France, Italy and Japan have increased the number of suppliers though they were created for various purposes including reducing foreign exchange costs and diversifying supplies. Supply contracts with state trading countries can have the effect of stimulating competition and of securing low energy prices although other considerations enter into these arrangements also.[1]

Finally, a large number of governments provide official assistance for research and development in the energy industries. There are, of course, special government organizations or agencies in many countries for new fuels, especially in the field of nuclear energy, such as the Atomic Energy Commission in the United States, the Atomic Energy Authority in the United Kingdom and the Commissariat à l'Énergie Atomique in France. All three are State financed and are heavy spenders of public money. But despite the undoubted desire of the British and French governments to boost their nuclear industries, both these countries, weighed down as they are by the burden of their international political aspirations, are not in a position to allocate more than a fixed proportion of their national income to their overall research and development programmes. While these figures may compare favourably with that of the United States in relative terms, in absolute terms they are, inevitably, rather modest. To

[1] *Ibid.*, pp. 117–18.

this, the relevant page in the O.E.C.D. Energy Commission's Report bears an eloquent testimony:

> The most impressive assistance towards research and development is that of the United States Government. The Federal authorities contribute to research bearing on all aspects of the discovery, production, conversion and use of solid and liquid fuels, to the study of certain aspects of energy transportation and to all sides of research and development of nuclear power, including construction of prototype nuclear reactors. According to an estimate made by the Bureau of Mines, out of 17 to 20 million dollars spent in 1959 on research relating to the production and use of coal, 10 million was provided by Federal authorities. Total research expenditure in the field of hydro-carbons was much higher; out of some 300 million dollars, the government provided about 10 per cent. In addition, almost 25 million dollars out of a total of 40 million dollars were contributed towards shale oil research. In total, well over a billion dollars has been spent for research and development in the field of nuclear power. Other countries too are spending large sums from government or public authority resources on research and development; for example the nationalized fuel industries in the United Kingdom are spending £15 million a year.[1]

The contrasting orders of magnitude of the two sets of figures hardly requires any comment.

Energy policy in the wider economic context

It is manifestly impossible to separate or compartmentalize energy policy from wider economic (and, occasionally, strategic and political) considerations. There are continuous intimate points of contact with financial, social and economic policy, such as helping to finance the national budget, balancing foreign payments, stimulating national economic growth, protecting other resource values (i.e. no land damage), and regional development and retraining of labour.

First, energy's contribution towards the financing of the national budget: this arises from the taxation of a wide spectrum of energy products for purely fiscal, as well as protective and other allied reasons. This is a subject we have already examined at some length, and there is probably little that we can usefully add, except to give an example of the importance of the contribution energy taxation makes to total government revenue. In the United Kingdom, in the financial year 1967/68, total government revenue from taxation amounted to £10,804 million. Of this, taxes on energy products contributed to less than £969 million or 9 per cent.

Secondly, there is the need to square the balance of payments. This, clearly, is a sensitive as well as a substantive point, and one of which all

[1] *Ibid.*, p. 119.

indigenous energy industries which have been forced onto the defensive are wont to make much play. It was estimated by the Commission of the European Communities in its latest forecasts that by 1980 the Community could well be dependent upon overseas suppliers for some 60 per cent of its total requirements. Quite apart from the strategic problems raised by a degree of dependence of this order of magnitude, which have been perhaps somewhat exaggerated in the past few years, there is a very real issue of cost. Thus, if we accept a figure of dependence upon overseas suppliers of 60 per cent, equal to a total energy requirement of 678 million tons, and an average delivered price of, say, $10 to $12 per ton of coal equivalent supplied, we obtain a total annual import bill for the Common Market countries of between $6,780 and $8,136 million a year. For the United Kingdom, a rundown of the British coal industry, at the rate envisaged in the 1967 White Paper on Fuel Policy, could mean an additional annual import requirement of over 100 million tons a year. These are enormous and, indeed, potentially critical sums of money. For countries like Germany in a strong and favourable creditor position, the risk of spending these sums of money in order to obtain cheaper energy and thereby release uneconomic factors of production at home is probably small, and a net financial gain on the outcome of the overall operation almost guaranteed. In a country like Britain, the decision to close down uneconomic mines, to redeploy the manpower released in this way in the most profitable manner to the national economy, and to obtain access to cheap energy for British industry, although no less vital and logical an objective, does involve a calculated financial risk—although one that must nonetheless be taken in the country's best long term economic interests. The governmental experts in the O.E.C.D. Energy Commission were no less convinced that this was the right path to pursue, even if temporary deviations from it could be tolerated in the short term:

wherever non-competitive indigenous production is protected on payments grounds alone, as is occasionally the case, the effects of such intervention become extremely difficult to assess. Obviously such protection has to be seen in the entire context of a country's international economic situation and competitive position, and in a long-term rather than in a short-term context. The benefits of protection must be weighed against the costs which it imposes on the economy. Thus, import restrictions on energy may lead to an increase in the price of energy within the country concerned which may prejudice the competitive position of its manufacturing industries, in particular of those in which energy costs are a relatively high proportion of total costs, such as the iron and steel industry. Therefore, the long-term solution to balance of payments problems would seem to lie instead either in making the indigenous industry competitive or encouraging the transfer of labour and capital resources to more productive sectors, in particular manufactures for export. Although that determines the general principle, special problems in certain countries, in particular countries in the course of development, may justify deviations in the short term.[1]

In short, the message in this inter-governmental document comes across loud and clear. The proper objectives are to reduce subsidies; to close down non-competitive energy sources and to balance rising import costs by lower energy costs for home industry. As a conclusion this is unexceptionable, except that the sum is not complete and that the evaluation of other factors, namely the impact on economic growth and the weight of regional and social policy considerations still remains to be done.

The significance and weight of these factors were clearly spelt out, though not quantified, in the Second Report from the U.K. Committee on Nationalized Industries in 1968 which stated that:

The true costs to the community of supplying the nation with the energy it needs is measured by the value of the resources employed—manpower, capital, material and foreign exchange. In their work on fuel policy the Ministry have been concerned to compare the possible alternative developments by reference to their relative costs in resource use. In general, the prices of fuel will reflect the cost of the resources used in supplying it. In certain circumstances, however, prices may diverge from true resource costs, for example:

(*a*) prices may include elements which are transfer payments, and not true resource costs—the oil tax and the royalty payable to the Exchequer on natural gas production are examples;

(*b*) when one fuel is displaced by another, part of the costs of the fuel displaced may be unavoidable (i.e. may not be saved) in the short term, such as the capital invested in existing assets;

(*c*) similarly, manpower liable to be displaced, for example from the coal industry, may be regarded as costing less in national terms than the wages actually paid if all the men cannot for some time be profitably re-employed, and in the short term the resource cost of coal from marginal collieries may thus be lower than the accounting costs of its production (conversely, if the men can be more profitably employed elsewhere, as should be the case in the longer term, there will be a saving in resources and a gain to the economy by their moving into more productive work).

These examples illustrate how money costs may sometimes diverge from resource costs. It is often difficult to put precise figures on the resource costs of a particular pattern of development, but it is usually possible to estimate the order of difference between alternative courses.[2]

We have already dwelt at length upon the importance of energy as a driving force in our modern technological society. Obvious and immediate

[1] O.E.C.D. Report, *op. cit.*, p. 122.

[2] Select Committee on Nationalized Industries, *Exploitation of North Sea Gas*, 1968, Appendix II (memorandum by the Ministry of Power), p. 194.

examples of its impact upon economic growth are many; for instance, the entire success of the Tennessee Valley Authority concept in the United States depended upon the availability of massive supplies of cheap energy; the same basic concept lies behind the industrial development plans for southern Italy, the industrialization of south-western France with natural gas from Lacq, the creation of an industrial structure in northern Holland with cheap natural gas from Groningen, and the prospects for the industrialization of East Anglia or Humberside with cheap natural gas from the North Sea.

Social factors are probably the most important of all. It would be monstrous to ask the miners, whose entire working lives have been spent in surroundings that even today are often an insult to human dignity, to pay the cost, on behalf of society, of the great switch-over to the new fuels and the new technologies. Their position, and the very real problems of regional economics and regional structure that arise with them, must be faced squarely and openly. Thus, if it is true that it is wishful thinking on the part of the coal industry to hope that it will, in the longer term, be able to compete on level terms with oil or nuclear power, the arguments for a close and unbiased study of its short or medium term potential contributions deserve, at the very least, and for reasons of enlightened self-interest, very careful consideration. The basic problem is one of how the total costs of the operation, namely the closure of pits plus the cost of regional development plus the extra fuel import bill, but minus the effect of cheaper energy costs and the results of the more rational utilization of available manpower, are to be assessed. This is not an apologia for the coal industry—far from it. Emotion has no place in the realm of economics, and the coal industry must face its future on every count: economic, regional, social and human. But there is an excessive tendency in many countries to write off coal as the fuel of a former age and rush headlong into the nuclear era without a clear notion or appreciation of the immense cost in money and people in acting in a manner that is at once precipitate and costly.

The fifth and last factor under this heading is that of protecting natural resources or amenities. This was a subject on which the O.E.C.D. Energy Commission had some strong words to say:

> the damage to land, water, fisheries, scenery and other resource values resulting from the production, the transport and the use of fuels (as well, of course, as other industrial operations) is beginning to come under close examination in many O.E.C.D. countries. Destruction of the land surface in mining, especially as the result of strip mining, but also through subsidence in underground

mining, damage to scenery by overhead electricity transmission lines, oil and gas drilling and the harnessing of mountain streams for the production of hydro-power, the pollution of groundwater, streams, lakes and seas and destruction of fish resulting from acid mine drainage, refuse products from refineries and tankers and pipe like leakage, added sediment, and increases of water temperature (thermal pollution) resulting from discharge of coolants from thermal generating plants are examples of the costly damage that may result from the production and transport of fuel and power.

The result has been growing government insistence on adequate guarantees of land restoration and industrial hygiene:

...legislation and regulations apply to virtually all sectors of energy production and transport, and far-reaching international co-operation exists in those fields where the hazards are greatest. Coal mining being one of the oldest forms of mass production of energy, it has already become traditional for governments to insist on the reparation of mining damage in coal producing countries. In the more recent cases of opencast brown coal mining in Germany, strict regulations exist to guarantee the restoration of the countryside. In the meantime, in some O.E.C.D. countries, the coal industries are making special efforts to minimize such damage. Much work has been done in restoring the surface in areas of opencast mining; it has been possible in places even above its original condition and hence to increase the value of the land. Similarly, recent experiments show that increases in water temperature can be turned to advantage locally in increasing fish production. As in many aspects of energy problems, scientific and technological advances can contribute greatly towards finding long term solutions.[1]

Nor is coal the only source of energy to be affected. In 1969 the Swedish Government took a series of measures designed to reduce the level of sulphur emission from oil-fired plants to 1 per cent in Stockholm and to 2½ per cent in country areas, with provisions for further reductions in the near future. Such measures, if applied generally throughout Europe and North America would involve both the coal and oil industries in very substantial desulphurization expenditure.

The objectives of energy policy

The objectives of a successful energy policy have been defined as follows:

(i) ensuring the security of energy supplies;

(ii) ensuring adequate energy supplies at the cheapest possible price;

(iii) ensuring the progressive substitution of new competitive fuels for the older traditional but less economic sources of energy;

(iv) ensuring the stability of energy supplies in terms of price as well as supply;

(v) maintaining a free choice of fuels for the consumer;

[1] O.E.C.D. Report, *op. cit.*, pp. 124–5.

(vi) maintaining fair conditions of competition between the competing fuels; and

(vii) fitting energy policy into the overall economic framework.

The virtues of these objectives are, of course, unassailable. The art of successful government lies in their successful permutation under whatever circumstances prevail at any given time. Thus, the British Government's White Paper on Fuel Policy of November 1967, after indicating the anticipated fuel pattern for the mid-1970s, concluded with the unexceptionable words that:

> the Government's aim has been to lay down long term policy guidelines which can be modified and adjusted in response to new developments and to provide a coherent framework within which specific decisions can be taken as the need for them arises. Within this context, future decisions can be based on up-to-date information; there will be periodical revision of the relevant economic analyses and estimates in the light of changed circumstances; and the Government will keep policy under continuing review.[1]

In an ideal world, the role of any government should be limited to that of an umpire, sitting inviolate in his high chair, and needing only rarely to intervene. Basically, underneath the camouflage of protective measures of various kinds, this is in fact what many governments have attempted to do. The power of the economic forces at work in the energy market is irresistible. Given the importance of energy and energy costs in the overall industrial structure, no government can for long pursue a dear energy policy that is grossly out of line with that of most of its rivals or competitors. Thus, once the Dutch Government had taken its decision in 1966 to close down its coal industry, the repercussions, like the ripples that quickly spread throughout a pond once the first stone has been thrown, have resounded throughout Western Europe, in Belgium, France, Germany and the United Kingdom with tremendous force and rapidity.

In the Common Market, the work of attempting to formulate a common, or at least a co-ordinated energy policy, has now been in progress since the late 1950s. Successive proposals met with little success, in part because of the divisions of power and responsibility among the three European Executives (i.e. the High Authority with responsibility for coal; the Euratom Commission with responsibility for nuclear power; and the European Commission with responsibility for oil and natural gas) but, primarily, because of the fundamental conflict of interests between the coal producing countries, in favour of a protectionist energy policy, and those countries with no indigenous energy industries, wholeheartedly in favour of a liberal energy import policy. Gradually, the economic impera-

[1] White Paper on Fuel Policy, *op. cit.*, p. 2.

tives of a cheap energy policy have come to be accepted by the governments of all six Member countries and this, together with the merging of the former three Executives into a single European Commission, has paved the way for a new attempt to develop a closer degree of co-ordination of the energy policies of the individual Member states. Before examining these most recent proposals, it is worth stressing the moral that this situation has come about, not as a result of hard bargaining leading to eventual agreement, with a certain amount of give and take among all the countries concerned, but because these countries have been compelled by force of economic circumstance to adopt energy policies which, while allowing for different rates of progress, are now basically similar in their purpose and objectives.

Towards the end of December 1968 the Commission approved for submission to the all-powerful Community Council of Ministers a report proposing new guidelines for a Community energy policy. This report, which was conceived as a basis for discussion by the Council of Ministers, first emphasized the need for a common energy policy as a means of achieving the old chestnut of a fully integrated common market and thus removing the dangers, implicit in the existing situation where Member states pursue their individual energy policies, of distortion of competition, uneconomic investments, and the fragmentation of the European energy economy. The need for an energy policy was also underlined by reference to the particularly important role played by energy in the Community's economic life and its dependence on imports for over half of its import requirements. Energy, it was recalled, represented today some 12 per cent of the Community's industrial production and 15 to 20 per cent of its investments. In 1967, moreover, the Community imported $5,500 million worth of energy products, i.e. about 18 per cent of its total imports. The report stated that the central principle of Community energy policy must be the protection of consumers' interests and that the main aim must be to provide assured supplies at prices which are both stable and as low as possible. It was once again recognized, however, that social and regional factors should influence the rate at which one source of energy replaces another, and the Commission promised that it would very soon propose measures concerning the social and employment problems of the coal industry.

The Commission stressed that the guiding force in the operation of the energy market must be competition, but argued that because of the extent of dependence on energy imports and the need to ensure long term security of supply, and having regard to the peculiar structural features of the energy sector, there had to be provisions enabling closer Community or joint observation and steering of energy developments than for most of the remainder of the economy. Direct intervention by the Commission should be envisaged only as a last resort, on the basis of recommendations by the Commission to the Council of Ministers in the event of market developments making such a step necessary.

The Commission proposed an 'action framework' for an energy policy which would be based on Community medium term forecasts and guidelines covering both the overall energy market and the individual forms of energy, and providing not only for the continuation of work already prescribed by the Coal and Steel Community and Euratom treaties on general objectives for coal and indicative or target programmes for nuclear energy, but also for similar arrangements for the electricity sector and medium term guidelines for oil and natural gas. In addition, the Community would prepare annual reports on the current energy situation and short term future outlook, on the basis of which the Commission would, if necessary, recommend adjustment measures to the Member states. To guard against the danger of supply difficulties, in particular that of interruption of imported supplies, the Community would also continuously examine the overall supply situation and implement policies for the stocking of oil and nuclear fuel.

On the setting up of a common energy market the Commission suggested measures for ensuring the free circulation of products and freedom of establishment, including the removal of technical obstacles to trade and of discrimination in such fields as oil prospecting, refining and distribution concessions. In the sphere of competition, the Commission emphasized the need for measures to ensure that consumers had access without discrimination to the sources of energy supply, and suggested that in view of the already small number of large concerns controlling oil, natural gas and nuclear energy in the Community, further mergers contemplated in these fields should be subject to prior notification to the Commission. A procedure should also be envisaged whereby the Commission would be kept informed *a posteriori* of prices effectively practised on the market for the various forms of energy; and the advisability should be considered of harmonizing at Community level the application of national pricing policies. The Commission further proposed that distortions of competition and consumer choice due to differences in indirect taxation of the various forms of energy within and between Member states should be eliminated by the harmonization of the added value taxation system in the energy sector and of consumption taxes on energy products, which should be progressively reduced when they are designed to protect particular forms of energy.

Finally, the Commission suggested a basis for establishing a policy for cheap and assured energy supplies. It proposed a co-ordinated import policy linked with Community guidance of the development of its own energy industries as part of an overall plan for Community energy supplies. This plan would enable the Community's institutions to assess whether there was sufficient diversification of sources of energy supply, according to types of energy, countries of origin and supplies, and whether sufficient regard was being paid to indigenous production recognized as important to the Community; it would also enable the Commission to make recommendations for adjustment measures such as temporary import restrictions and sales guarantees should the Community's interests require them. Provision should be made for the Commission to be informed of important investment projects affecting energy in the fields of production, transport and distribution. The Commission and representatives of the Member states should jointly discuss investment plans and policies annually in the light of the medium term energy forecasts, and proposals contrary to the Community interest which could not be adjusted by these means should, if necessary, be dealt with by formal Commission recommendations to the Member states concerned. In this connection attention was drawn to the need for a rational strategy for the construction and development of nuclear reactors, the co-ordination of power station projects to secure optimum location and efficiency from the Community point of view, and the overall perspective of oil refinery projects in order to avoid over-investment. The Commission also suggested that an arrangement should be made for Community financial assistance for investment and research projects contributing towards the aims of Community energy policy.

On the development of the structure of the Community energy industries the Commission stated that the objective must be to shape the energy sector in such a way that the rapidly growing demand for energy could be met on the proposed basis of low costs and security. The structure must be so influenced as to ensure the maintenance of healthy competition; at the same time appropriate means must be found for promoting desirable amalgamations of, and co-operation among, Community enterprises, including firms in differing countries of the Community. As regards coal, it was pointed out that the existing production and distribution structure could in certain instances prejudice the success of the efforts which the coal industry is making to rationalize production and improve costs in order to create a balanced coal economy. The Commission's main proposal on coal was that the medium term coal production plans of the Community countries concerned should be examined as a whole, taking into account the need to adjust Community output to market demands and to concentrate the remaining output on the pits with the highest productivity. Regard should be paid to the position and location of pits in relation to markets and to their reserves of the qualities of coal in demand. The economic and social development of the regions affected must also be considered. The Commission also suggested that a Community aid system should be introduced to enable a level of output commensurate with the overall balance of the Community's energy supply to be achieved; that the existing Community aid systems and the individual national measures of assistance to the coal industry should be better co-ordinated and linked with the attainment of the Community's energy policy objectives; and that measures should be taken to allow for the retention and recruitment of the manpower required for the realization of the Community's energy policy.

With regard to oil and natural gas, it was suggested that steps should be taken to remedy, by means of appropriate fiscal measures, any disadvantages suffered by Community oil companies in meeting the competition of the main third country big international oil companies, having their headquarters outside the Community; to facilitate the acquisition by Community oil companies of overseas crude oil concessions; and to promote the formation of larger Community oil enterprises, by means of mergers or other measures such as the rapid elaboration of a European company statute. In the field of nuclear energy, the Commission urged the construction of a European isotope separation (i.e. uranium enrichment) plant; and the encouragement of prospection for uranium by Community enterprises both within and without the Community, again, where appropriate, by means of the development of a European statute. Lastly, the Commission recommended a programme of co-ordination of research in the oil sector, and the promotion of scientific and technical research into new techniques of energy production, as well as with a view to finding a solution to the pressing problems of water and air pollution.

The Commission's underlying task was seen as being to act as a powerhouse for the continued expansion and development of the Common Market on Community principles. Its proposals in the energy field may be said to be applicable, in the broad term, to almost any major industrial power. Nevertheless, it is illustrative and instructive to see how difficult it has been, and continues to be, for these six countries, irrevocably committed though they are to the cause of union and progressive economic integration, to reach agreement in a field where their fundamental interests, however contradictory they may have been initially, have, for the past few years, been rapidly converging.

Nor is it only in Western Europe that pressures have been growing for a common or at least a co-ordinated energy policy. We have seen already how the often conflicting aims of North American oil producers, consumers and strategists are all pressing, in one way or another, for a clearly defined continental policy for North America. In a situation where the producers are anxious to retain their favourable American market conditions—free from meaningful competition from cheap Middle East oil and able to notch up safe and high profit margins; where consumers, on the other hand, are anxious to achieve as high an import quota as possible, and even complete relaxation of oil imports if ever the opportunity should arise; where the U.S. Government is anxious to maintain a high indigenous output and reserve ratio and yet must, at the same time, keep room, for political reasons, for Canadian and Venezuelan oil imports; and where other governments, notably that of Canada, are pressing hard for higher oil exports to the United States and for a long term North American continental oil policy, the prospects of finding a policy that will please all can scarcely be said to be bright.

The whole complex problem of a North American oil policy was well summed up by a Canadian commentator writing in *Oilweek* early in 1969:

> Concern over protection of Canadian oil markets in the United States arises from fear that prolific oil reserves on the Alaskan north slope will displace U.S. sales of Canadian oil. Bolstered by a hefty increase in its reserves position, there is some apprehension that protectionist influences in the United States will seek more stringent curbs on imports of Canadian oil.
>
> Alarmists who fear that Canadian oil would be pushed out of the U.S. market reveal only half the picture. Certainly there is at least some risk, inevitably will we lose markets on the U.S. west coast, and some protectionist U.S. sentiment would like to see this happen in the large U.S. midwest market.
>
> But there are also U.S. interests with billions of dollars invested in Canadian oil who would like to see a better market for reserves they have developed. There are substantial U.S. refinery interests which would like much freer access to Canadian crude. And there are strong consumer interests which would welcome such freer access to economic crude supplies.
>
> There is a strong U.S. demand for Canadian oil, and for one primary reason: it represents a convenient, dependable and economic supply source. And there is no reason why Canadian oil should not be able to maintain this position.
>
> The North American petroleum industry is based on a high price structure, relative to other world crude prices. This is the cost of defense security and supply assurance—and it is well worth a premium.
>
> But there is a limit to the premium that will be paid before the entire structure is shaken. Intense pressures for greatly expanded refinery capacity on the

eastern seaboards of both Canada and the United States, based on low-cost imported crudes, indicates we are very near that limit.

If North America is to maintain adequate petroleum reserves and supply the bulk of its own requirements, it must utilize the continent's petroleum resources in the most economic manner possible. And this means a common market in petroleum.

Consider the alternatives. The United States sets a policy to displace as much Canadian oil imports with Alaskan crude as possible, regardless of costs. To do so means moving Alaskan crude into the U.S. midwest market, not by the most direct and economic manner—a pipeline from the north slope stretching across Canada—but by much lengthier combination tanker and pipeline routes half way around and across the continent. The added transportation costs may be between half a million and a million dollars a day. Pushed out of the closer and more economic U.S. midwest market, Canada would likely extend its markets for western oil to Quebec and the Atlantic seaboard to displace imports, again adding to costs.

The structure of protection would build higher and higher until it inevitably falls with a horrendous crash and devastating effects. Sooner or later this would be the certain result.

Initially at least, oil would be the most important factor in a North American energy market. But coal, natural gas, hydro-electric power and uranium are also vital.

Establishment of a North American energy market would add billions of dollars to the Canadian economy during the next decade. It would mean no less to the United States. North America's potential resources in likely new petroleum provinces—Alaska, the Canadian Arctic and offshore—are enormous. To harness this potential represents an equally tremendous economic challenge which will demand a new dimension in Canadian–U.S. co-operation.[1]

Growing similarities in energy policies of all the major industrialized countries in the shape of an unmitigated recognition of the virtues and advantages of a cheap energy policy and, consequently, a greater readiness for, and the gradual dawn of, a better climate for co-operation in the energy field are, therefore, likely developments for the short or medium term future. As we go, cautiously but nonetheless steadily, towards this goal, let us recapitulate for a moment on the overall effect of current policies.

The first objective of obtaining adequate supplies of energy at relatively satisfactory, if by no means ultra-cheap prices, has been largely achieved, mainly as a result of the initiatives taken by the oil industry, backed up by government determination in a growing number of countries to seek to reduce the cost of energy to industry. The visible outward forms have been the rapid growth of refineries, electricity networks, pipelines and giant tankers. As we have seen, a very large element of protection for uneconomic indigenous industries remains and is likely to continue to do so for some years to come. It is significant nonetheless that the long term goal of cheap

[1] *Oilweek*, 10 March 1969.

energy has now been widely recognized and that a rapidly growing number of countries are taking urgent measures to accelerate its realization. Their effect on other more hesitant countries is bound to be cumulative. For the present, however, these protective policies undoubtedly contribute to the maintenance of a relatively higher level of prices:

...measures of protection for indigenous energy add to the cost of energy in the short term, even through they may be designed to bring about the lowest practicable cost to the community in the long term. But the effects on costs and prices are complex. Tariffs and import restrictions clearly increase the price of energy to the consumer in the country concerned; but an interesting side effect of the United States oil restrictions is that they have probably tended to reduce world oil prices and hence the price paid by other O.E.C.D. countries. Subsidies and tax concessions granted to non-competitive energy industries may keep down prices and those given to consumers may compensate for price increases arising from protective measures; but, like other forms of protection, they are a burden on the community. Taxes and other duties normally increase the price to the consumer to whom they are passed on, although at times (as is thought to be the case with some fuel oil taxes) they may be partly or sometimes wholly absorbed by producers or dealers. Because of these complications, it is particularly difficult to quantify the effects of such protective measures. Any estimate of the effects of government policies on energy costs to the community is further complicated by the difficulty of putting a figure to the social and economic costs and benefits involved in different policies or of the long-term effects on costs and prices of measures taken to improve security of supplies.[1]

Few countries today can hope to find quick and easy solutions to their energy problems. Those that can, like, for example, Italy, with no high-cost indigenous production and an accepted policy of importing energy from the cheapest possible source, largely regardless of the type of political regime in power in the supplying countries, are in a fortunate and, among industrialized countries at least, in rather an exceptional position. For most other countries and governments, it all boils down to finding the most satisfactory permutation among the various, universally accepted objectives of energy policy on an international or, where no other solution exists, on a national scale. We have already had occasion to look at the British Government's White Paper on Fuel Policy in Britain up to the mid 1970s for an example of energy policy on a national level, and at the efforts that have been made by the six Common Market countries to achieve the co-ordination of their energy policies as an example of energy policy on an international scale. The proposals aiming at a North American continental energy policy are evidence that the trend is not limited to Europe. Other proposals for co-ordination of energy policy, less precise and very much less ambitious in nature, were also put forward by the

[1] O.E.C.D. Report, *op. cit.*, pp. 128–9.

Energy Commission of O.E.C.D. Their basic premise was that all Member countries would stand to gain from their joint recognition of a broad framework of economic and social objectives, consisting of:

(*a*) the realization in the long term of the lowest energy costs for the community as a whole and of reasonable and stable prices for each consumer;

(*b*) secure and regular supplies (i.e., publication of energy prices and fuel transport rates);

(*c*) competition and transparency in energy markets;

(*d*) consumer freedom of choice between competing fuels on the basis of prices which accurately and fairly reflect costs of supply;

(*e*) safeguarding public health and preserving other resources.

Where, it was added, subsidies or protection were necessary, they should be regarded as temporary, and limited to measures designed to achieve rationalization of efficient producers. At the same time governments should carry out assessments of the advantages and disadvantages of adding another burden to the public exchequer and of the sum cost to the community, including the effect of regional and employment problems; they should also be prepared to initiate rapid changes of policy in the light of improvements or, alternatively, serious setbacks in the competitive position of the protected sector in question.

Against this background, all Member states of O.E.C.D. were invited to accept the following, deliberately loose, but nonetheless valuable proposals for international collaboration in the energy field:

(*a*) co-operation in attaining policy objectives common to all or a number of Member countries;
(*b*) confrontations and discussions on measures of national policy, particularly those which can affect other countries directly or indirectly;
(*c*) encouragement of the optimum use of available resources by means of international technical co-operation.

Since this O.E.C.D. Report was published in 1966, a good ideal of progress has in fact been made in the forum of the Energy Commission, particularly under the head of (*b*). Confrontation meetings have been held on coal problems; on energy policy; on the fiscal burden of the various forms of energy; on air pollution resulting from the production and use of fuels; and on the national administrative structures for dealing with energy policy and problems. In addition, there have been a number of 'by-country' confrontations, at which representatives from individual countries are invited to submit papers setting out the main imperatives of their countries' energy policies and to join in a probing discussion on their possible merits or disadvantages. The essential crux of the O.E.C.D.'s contention was that, in energy as in all other fields, all countries are today interrelated; this is true of industrialized and non-industrialized countries, of oil-producing as of oil-consuming states. As the Report concluded in one key sentence: 'While the situation of both groups of countries [i.e. energy exporters and energy importers] will change and the pattern of trade will alter, the essential inter-dependence will remain.'

Conclusion

The basic fact, truth or contention (the precise word chosen depending upon one's point of view) of the situation is that far from there being any indication of an imminent fuel shortage, all the signs point to a continuing and indeed an increasing abundance of energy supplies as far ahead as it is possible to see at the present time. In these circumstances, the crucial tasks of energy policy, both at present and in the immediate future, consist of finding a reasonable and acceptable balance between different sources of energy, some old and traditional, some new and dynamic, and of obtaining them at the cheapest possible prices within the framework of the wider economic, social and regional considerations that are the very stuff of government. But these unexceptionable objectives are, as always in such cases, very much easier to define than to achieve.

Energy problems and, indeed, energy crises have been a familiar problem since the end of the Second World War. For the first ten or twelve years of the post-war period there was an energy shortage with problems that were sometimes acute and alarming of ensuring an adequate energy supply. The situation changed dramatically towards the end of the 1950s and gave way to an energy surplus which has reached almost embarrassing proportions. This in turn has led to the coal crisis which is still very much with us today. It is coal, as the least economic and the most inflexible of all the major sources of energy, that has had to bear the cost of this revolution in energy supplies. There is no doubt that this problem in its turn will fade, whether it be as a result of the virtual disappearance of coal production or, less realistically, a remarkable breakthrough in coal mining productivity. But of one thing we may be sure, that the coal problem will not be the last in the field of energy. The shape of some future problems can already be discerned on the horizon: such as the very real problems and dangers of atmospheric or water pollution and the ways in which they can be overcome; or the future conflict between nuclear power and the hydrocarbon fuels, particularly in the power station markets of the industrialized countries of the world. At the same time, although the future battle between nuclear power and oil seems hardly likely to come to a head before the 1980s or even, perhaps, the 1990s when the fast breeder reactors come into their own, and long also before the fuel cell shows signs of becoming an immediately active and commercial reality, the oil industry is already busily expanding into a number of sideline activities, of which petrochemicals are by far the most important. It has been estimated that by the year 2000 petrochemicals may account for as much as 35 per cent of all oil consumption, compared with only 2 per cent in 1965 and an estimated figure of 10 per cent by 1985. The extraction of food products is another potential market for oil of immense significance. It should not be overlooked, for example, that as long ago as June 1967 one leading oil company issued a statement in which it said that its research workers were engaged on developing a process in which yeast-like organisms are nourished by high-purity hydrocarbons. This process, it was claimed, had 'already proved successful in yielding a food source greater in nutritional value than the proteins contained in most cereals and vegetables and substantially equal to the protein contained in meat, fish, milk and eggs'. Since even further diversification by the international oil industry is highly likely, the conflict between oil and nuclear power, if and when it comes to a head, seems hardly likely to generate problems of such human and social intensity as those we associate today with coal.

The answer, unsatisfactory though it may seem in its incompleteness, is one of continuous review of the changing kaleidoscopic conditions on the economic, social, scientific and technical fronts. In many of these fields it is utterly impossible, as well as foolhardy, to attempt to make accurate detailed forecasts of developments more than five, or at the most ten years ahead. This dynamic situation makes continuity of review and a readiness at any time to reconsider some of the sacred cows of earlier generations of official thinking a *sine qua non* of a successful energy and, indeed, overall industrial policy. Nevertheless, when all has been said and done, the basic essence of the situation is such that there is no longer any clear reason why the governments of all the industrialized countries of the western world should not formally adopt as a first principle of their energy policy the firm decision to go ahead, in the best interests of peoples everywhere, and whatever their field of activity or interest, be it commercial, industrial or domestic, as quickly as their various and varying social, regional and human obligations will allow, towards an avowed, dynamic and undisguised cheap energy policy.

INDEX